Wall Street Meat

Wall Street Meat

My Narrow Escape from the Stock Market Grinder

ANDY KESSLER

HarperBusiness
An Imprint of HarperCollins*Publishers*

HarperCollins books may be purchased for educational, business, or sales promotional use. For information please write: Special Markets Department, HarperCollins Publishers Inc., 10 East 53rd Street, New York, NY 10022.

First HarperBusiness paperback edition published 2004

Designed by Amy Hill

Library of Congress Cataloguing-in-Publication Data
Kessler, Andy.
Wall Street meat : my narrow escape from the stock market grinder / by Andy Kessler.—1st HarperBusiness paperback ed.
p. cm.
Includes index.
ISBN 0-06-059214-1 (alk. paper)
1. Wall Street—Anecdotes. 2. Speculation—United States—Anecdotes. I. Title.

HG4572.K47 2004
332.64'273—dc22

2003057074

04 05 06 ❖/RRD 10 9 8 7 6 5 4 3 2 1

For Nancy and the Boys

Contents

Foreword by Michael Lewis

Jack Grubman, Frank Quattrone, Mary Meeker, and Henry Blodget, were Wall Street's best-known promoters of the Internet-telecom boom. A few years ago, they were spoken of in a single breath, and in flattering tones, but those times now feel as ancient as Mesopotamia.

For more than a year, New York Attorney General Eliot Spitzer has made as good a political living off the three men as has ever been made on Wall Street. All three are meeting grisly ends.

Spitzer rooted out and publicized a few of Blodget's old emails, in a way that anyone who followed Blodget's career knew badly distorted the man's motives and character. Now Blodget is the subject of a NASD probe that may lead to a lifetime ban from the securities industry and a multimillion-dollar fine.

Grubman, the Caliban of Wall Street, has been fined $15 million and banished not merely from the securities industry but from New York City's private kindergartens, too. And while Quattrone was charged in March 2003 with obstructing a probe of Credit Suisse First Boston, he faces the threat of penalties even worse than Grubman's.

Meeker is the exception. She has not been fined, banned, or even asked by her employer, Morgan Stanley, to take a paid vacation. Spitzer seems to have abandoned whatever interest he had in ruining her career, or in singling out Morgan Stanley for its role in the boom.

And Morgan Stanley's executives have clearly decided to stick by their most famous analyst. In early April, at the Morgan Stanley annual meeting, shareholder advocate Evelyn Y. Davis, a character made to order for some Hollywood screenwriter, used Meeker as a club to beat Morgan Stanley Chief Executive Officer Phil Purcell.

Davis pointed out that "a lot of people lost money because of her. She should be fired." But while Purcell wilted beneath other blows from the 73-year-old Davis, he stood right up to her on the subject of Meeker. "We have a very different view of the contribution Mary made," he replied. "She was a pioneer on the Internet. We value her research."

It's a curious situation, and it becomes a great deal more curious when you read this deliciously naughty book, *Wall Street Meat*.

Andy Kessler is a fellow I've known for some time, but I had no idea he was writing a book. One dark day the thing just showed up on my desk, and I breathed a heavy sigh—another goddamn friend had gone and written another goddamn book, and I was going to have to at least pretend to read the goddamn thing.

To slake my conscience, I picked it up and read a few pages. My heart sank. It wasn't bad enough to put down and lie about having read it. I read a few more pages . . . and a few more. I finished it in a gulp, perfectly astonished.

Kessler, a former Wall Street technology analyst, worked with Grubman at Paine Webber Group Inc. and Meeker and Quattrone at Morgan Stanley. He finally quit in the 1990s to make a fortune putting money where his mouth was, actually investing in technology companies. He was right there, on the inside. He knew exactly how the Wall Street machine worked.

Having made his fortune, he now seems to feel free to say what he wants about his former firms and old colleagues. What he has to say will not particularly please them, but that is neither here nor there. It will interest the rest of us.

Wall Street Meat shows what the essential problem of Wall Street research was—as most people now know—that it ceased to serve investors and began to serve investment bankers.

Investors no longer paid the investment banks well enough so that they could afford to produce honest research, and so the investment banks figured out how to use their research to help the corporations that paid them far better. And the investment bank that forged this dubious new model was . . . Morgan Stanley.

When Kessler quit Paine Webber to join Morgan Stanley, he only moved about a hundred yards down the street. But in that little distance, he traveled from an old and dying Wall Street to a new and thriving one.

At Morgan Stanley, Kessler learned that he would be paid not by the brokers but by the bankers—and that the banker who would make the most difference to him was Quattrone.

Quattrone soon would leave Morgan Stanley, first for Deutsche Bank and then for Credit Suisse First Boston, but not before he found and created an analyst who, in Kessler's view, was a) desperate to be one of the boys, b) free of useful investment ideas, and c) willing to define her job as the pleasing of corporations that issued securities rather than investors who bought them. That analyst was Meeker.

To Kessler, it seems clear that Meeker was particularly susceptible to her investment bankers because she had nothing interesting to say to her investors.

There are a lot of theories to explain why Spitzer let Meeker be. One is that it doesn't pay for a politician to beat up women in public. Another is that Morgan Stanley's tech people somehow erased most of their old emails, and so Spitzer was unable to humiliate the firm with its own watercooler chat. (CSFB, which was far less central than Morgan Stanley to the Internet boom, wound up coughing up 10 times more email to Spitzer's office.) Or perhaps Spitzer found nothing to suggest that in promoting the likes of Women.com Networks Inc. and HomeGrocer.com, Meeker ever had anything other than investors' interests at heart.

Regardless of which theory you believe, Kessler makes one thing clear: Morgan Stanley, and Meeker, knew no other way of doing business.

In his view, Morgan Stanley, having no tradition of pleasing investors, and no ability to do so, designed its investment bank without the investor in mind. Meeker and her bosses didn't know enough to have a guilty conscience when they sold out investors to please corporations. In a game like this, that's a huge advantage.

Wall Street Meat

Introduction

My partner Fred and I are standing outside of Il Fornaio Restaurant in Palo Alto, the epicenter of Silicon Valley. It's July 1999, and although it's only eight in the morning, we have already been scrambling at the office for an hour and a half. It's a busy day, way too busy to be loitering around. Things are hopping. Internet and telecom names are jumping off the charts, so all my old friends are busy too. Jack Grubman keeps riding Worldcom to new highs—I think I saw it crack $65 this morning. Deals galore flow from Frank Quattrone at CS First Boston. Mary Meeker at Morgan Stanley discovers an Internet company a week to feel good about. It's hard to find time to even think.

We are doing someone back East a favor by showing up for this meeting. I'm just not sure with whom, let alone what it was all about. It's not a good day for it.

"Now, who is it we are meeting with?" I ask.

"I'm not sure," Fred responds. "I was just told to please meet with these folks and hear them out."

On cue, a big, black stretch limo pulls up in front of the restaurant. It is one of those extra stretch jobs, cut open in the middle and with a few more slices of limo added. Out of the back pour eight guys in suits. Three of them are the size of defensive lineman Warren Sapp and in ill-fitting suits, reminding me of Odd Job from the James Bond movie *Goldfinger.*

"Mr. Kittler? Mr. Kessler?" Well, here are our mystery guests.

We take a table for ten at the back. As we head through the restaurant, every table is taken and buzzing. I can't help noticing all the players sipping lattes. There is Jim Breyer from Accel Ventures, Steve Baloff from Advanced Technology Ventures. I think I recognize the CFO from JDS Uniphase. Isn't that the guy who runs Excite@Home? Hey, and John Sculley. "Good morning, John."

Our guests all speak fluent English. Over a round of espressos, they explain that they are from Bahrain and thank us for taking the time to meet. Of course, I finally realize, the three beefcakes are bodyguards.

One guy seems to be the leader and says, "We have heard about you guys and your numbers, and we are very interested."

Fred and I explain in a little more detail about what we do and what our style is. After a few minutes, the head of the group interrupts, "I like what I am hearing and am very interested. I would like to propose that we wire you $500 million tomorrow morning."

I think I pass out for a second, but as I come to, I hear Fred thanking them for coming all this way and saying that it was great to meet them. I try to kick him under the table but the tree trunk legs of one of the bodyguards are in the way. I can't believe it, $500 million and Fred is saying goodbye? That kind of dough is hard to scrape up. They pick up the tab, we walk them back out to their stretch-stretch limo, thank them for their interest and off they go.

"Fred, you just usher them out like that? That's a ton of dough. My kids need new shoes," I say in a slightly hysterical tone.

"I know. It is tempting. But if we take their money, we will end up working for them. I don't know about you, but I didn't get into all this to work for those guys."

"Yeah, but Fred. Think about it. There's all sorts of stuff going on that we could do with that much capital. Money is money, it doesn't matter where it's from."

"Sure it does, and there is too much of it floating around these days," Fred points out.

I think about all this and stew all the way back to our office. About an hour later, after sitting quietly for a while, I break the silence, "Dammit, Fred. I hate it when you're right. It annoys the shit out of me."

• • •

It's not every day that someone throws $500 million at you, but things get even stranger. A week later, another friend calls and insists we meet with someone at Il Fornaio the next morning for breakfast. So here we stand again, 8 a.m. at the same entrance to the same restaurant on Cowper, waiting for another set of mystery guests. This time, a white, normal-sized stretch limo pulls up. Four guys pile out. No bodyguards this time.

We sit at the same table at the back. On the way through, we see venture guys from Sequoia Ventures, Draper Fischer, and Mayfield. I recognize a few company managers from Avanex and Inktomi.

Our guests explain that they are from Saudi Arabia and thank us for taking the time to meet. It's kind of spooky. The head guy speaks up, "We have heard about you guys and your numbers and we are very interested."

I steal a quick glance at Fred, who rolls his eyes to let me know that he is thinking the same thing I am. We are sitting through the exact same movie as last week. Identical script. It is very creepy. To be polite, we start to explain what we do and our style, but we don't get too far.

With a wave of his hand, the head guy interrupts and says, "We know the story. We are very interested. I would like to propose that we wire you $500 million tomorrow morning." It seems that $500 million is some kind of petrodollar unit.

I don't pass out this time, but quickly respond, using the same words Fred had the week before. We quickly run out of the restaurant. On the way back to our dumpy office above an art store, I chime in, "You know, you haven't lived until you've had $1 billion thrown at you in one week."

Fred just chuckles and shakes his head.

If we don't take the money, someone else will. I try to think this through, but my head is spinning. Instead, all I can think of is what a strange trip it's been, since I got sucked into this bizarre world.

CHAPTER 1

Right 51% of the Time

I dialed the phone number in the ad.

"Hello, Research," said the woman on the other end of the line.

"Huh???" I thought. "Research?"

"If you could be so kind, may I speak with Robert Cornell, please?" I can turn on polite in a hurry.

"He's gone for the day. May I take a message?"

"Have him call Andy Kessler at this number, and by the way, research where?"

"Paine Webber."

Now I had seen the Thank You Paine Webber commercials with Jimmy Connors but had no clue what Paine Webber even was. No problem, I'll figure it out tomorrow. It's 1985 and I'm 26 and in no rush.

Bob called back the next morning. He had a deep exotic, take-charge phone voice. He grilled me on my background, and then asked when we could meet. I was headed into Manhattan from my house in New Jersey to meet with a headhunter who placed programmers and tech people in temporary assignments. I could meet for lunch.

I decided to wear the best clothes I had. A maroon shirt, navy blue wool tie and double knit Haggar slacks. No jacket. It was very early Geek Chic, circa 1985. I ran into a college friend on 51st and 6th Avenue, who asked me where I was headed. "An interview at Paine Webber." He gave me one of those you-dumb-shit looks. "Dressed like that?"

OK, so I was a dumb-shit, and now a self-consciously underdressed one. At 1285 Avenue of the Americas, I took the elevator up to the ninth floor and headed to the receptionist. Behind me, a door opened briefly. I caught a glimpse of a whole roomful of people who were yelling at each other and into phones. What a strange place. The reception area was filled with the worst art I had ever seen, ugly contemporary pieces that had probably been drawn by three-year-olds.

"Hi, I'm Bob Cornell," the surprisingly short and smallish Bob Cornell said, with that same deep phone voice I had heard the day before. As he pumped my hand, his eyes slowly looked me over from my face down to my shoes (did I mention my ugly brown Florsheims?) and back up. He had this funny smile on his face that I would later realize was an I-think-I-just-found-what-I was-looking-for look.

"C'mon in, let me introduce you to a couple of guys who are joining us for lunch. Steve, Jack, let's go."

A somewhat round, well-dressed, light-haired guy came over first and said hello with a British accent. "This is Steve Smith, he is the number one computer analyst on Wall Street."

The next guy was wearing a three-piece suit, with jet-black

hair combed back and a pointed, almost sinister-looking beard. He said hello with a Philly accent, an almost Rocky Balboa-like "yo." "This is Jack Grubman, our hot telecom analyst. Like you, he used to work at AT&T."

It was the start of a very wild ride.

• • •

Bob Cornell moved things along. "Let's head over to Ben Benson's. It's across the street and we can talk there." The four of us headed down the elevator and crossed 52nd Street. As we queued up at the Maitre d's station, Bob Cornell remarked to chuckles from Steve and Jack, "I hope they don't have a dress code in here."

We gorged on Flintstone brontosaurus burger-sized steaks. I had just spent the last five years of my working life at Bell Labs, the research arm of AT&T, spending huge amounts of ratepayers' money. I had designed chips, written lots of software for graphics workstations, installed million-dollar minicomputer systems. But AT&T was in the midst of maximum change. The breakup took place in 1982. I needed out.

I went through my electrical engineering background, the stuff I did at Bell Labs, and what I thought about computers and PCs and modems and fiber optics. It was time to go in for the kill. "I'm not really the corporate type. What you guys really need is a consultant like me." I meant to say it as a statement but ended up with the inflection of a question.

Bob jumped in, "We don't need no stinkin' consultants. We need an analyst, someone to follow the semiconductor industry." I wasn't sure what he meant.

I spent the next two hours telling them that I wasn't kidding; I really wasn't the corporate type. I didn't even own a suit, didn't know a thing about financials, only about technology. I must have struck a nerve. All three launched into a coordinated attack, convincing me to take the job as an analyst, a job I didn't even know existed until that morning.

The more I protested, the more adamant they were. It was a great negotiating technique. I wish I had thought of it ahead of time and done it intentionally, but I really was clueless. I occasionally wake up in a cold sweat worrying about the consequences had I been more convincing as a slacker that day.

Wall Street is filled with analysts, covering every imaginable industry. Oil analysts, retail analysts, beverage container analysts, auto analysts, insurance analysts, ad nauseum. A rip-roaring bull market had started in August 1982, and the hot stocks were those in technology—computers, semiconductors, and telecom. The old-line analysts who had been around for ten years waiting for this bull market to start were experts at IBM but not much else. IBM was half the computer industry's sales and 90% of their profits. But the bull market brought with it a boom in initial public offerings, IPOs, in 1983.

A bunch of weird companies were now public and Wall Street needed analysts to figure them out and recommend the best ones for investment. Those old dog IBM analysts were not equipped to deal with such things as design automation, PC software, gate array semiconductors, competitive long distance, workstations, disk drives—Wall Street firms needed smart analysts in narrower and narrower sectors to take public even more of these companies the next time an IPO boom came along. My technical background, proved by my fashion sense, was just what was needed.

Bob had put together a technology research team for computers, software, and telecom. I was the final piece of his puzzle.

I still wasn't convinced, until Bob whispered in my ear what I would make in the first year. It was about three times what I was paid at Bell Labs. I stopped protesting so loudly. He then said "And that's for being right just 51% of the time."

I made a few calls to friends of friends who knew about Wall Street, asking them what an analyst is and what they did. I didn't get very informative answers. One old college friend from a wealthy family who knew a bit about finance was direct. "What are you, some kind of stupid idiot? Take the job." OK, that worked.

I got the offer. My final act before entering the wild and woolly world of Wall Street was a bit of sly negotiation. I wasn't born yesterday—I was going to get something out of these folks. When I left Bell Labs, I was getting two weeks vacation a year and if I had stayed another four or five years, it would automatically have been bumped to three weeks vacation. I told Bob Cornell I was getting three weeks vacation a year at my last job, and I wanted three weeks vacation at this one. He paused, giving me the strangest look (ha, I've got him) and then told me, "Sure, I can write that down, but there is no set vacation time on Wall Street. You take as much time off as you can spare." These were prophetic words, which, of course, I completely misread. I figured I'd be taking ten weeks off each year, but for the next fifteen years, I barely squeezed out two.

My maroon shirt and wool tie were not going to cut it. Neither was my light blue leisure suit left over from college. I ran down to Mo Ginsburg's and dropped $1000, which was almost all the money I had, on four three-piece suits (the Jack Grubman look), and some shirts and ties. I was told yellow ties were hot. "Power ties," the salesman told me. Paisley was in, too. "Do you need braces—you know, suspenders?" I avoided them all. Red ties and a brown belt were about as racy as I was going to get.

• • •

On my first morning, I struggled to get into midtown by 8:15 a.m. I was told that there was a meeting every day at that time. I called up to Bob Cornell from the lobby and he told me I had to go to Personnel to check in, and fill out forms before I was allowed to enter the realm of research.

I met up with a pleasant, ferociously gum-chewing woman named Candy in Human Resources.

"What program are you in?"

"Uh, I don't think I'm in a program."

"Oh sure, honey, everybody is."

"Well, I can ask."

"No matter," she said as she inked my fingertips and rolled them on an FBI fingerprint form. "Ever been convicted of forgery?"

"No."

"Well, we'll check," Candy went on, "but you're in luck. Wall Street has very strict rules. You can be arrested and convicted of just about anything and still work on the Street. Anything but forgery, that is. There are too many documents and certificates lying around. Wouldn't work."

"I'll keep that in mind."

I finally met up with Bob Cornell around 10:30 that morning. He ushered me into my new office. It had a cherry wood desk and return, with matching filing cabinets and credenza.

There was also a leather chair for me, and two beautiful cherry wood chairs, with upholstery that matched the classy light green carpeting. A window looked out onto 51st Street.

A secretary I'd be sharing with Jack Grubman poked her head in and said, "Mr. Kessler, let me know when you will be taking calls." That was a little bit too much for a 26 year old to handle.

Bob left to do whatever it was he did. Jack Grubman, ear to a phone, stuck his head around my door from his office right next-door and nodded hello. Steve Smith strolled by and said, "Let's have a pint sometime." A guy named Jonathon Fram, who followed big computer companies, walked in and asked if I knew anything about Emitter Coupled Logic used in IBM mainframes. I did, but he was out the door before I could tell him. Next, Curt Monash plopped into one of my matching chairs. Curt graduated with a PhD at age twelve, or some such thing, had an IQ over 200, and was the top software analyst on the Street because no one else understood the intricacies of software as well as he did. How did I know all this? Curt told me. I mentioned that I had written some code and Curt went cold, as though I had called his bluff. I think I had.

. . .

At a quarter to twelve, the office was quiet. A head popped into view. "Got any lunch plans?"

"No," I said.

"Let's go."

It was Andrew Wallach, a cable TV analyst. He took me down to a cafeteria in the building. Over a stale turkey sandwich, he launched into me.

"I hope you go by Andy, because I go by Andrew."

"Sure, because, . . ."

"You know, I've been an analyst for about a year now. It's OK, pretty good actually, lots of leverage, you meet interesting people. It's hard work but what isn't." Blah blah blah, the guy just babbled, but ended with, "I really think that this is just a stepping-stone to something else down the road."

I've been working as an analyst on Wall Street for less than four hours, still clueless about what that even means, and already I'm being lectured that it is a stepping-stone to something else. Great.

"Uh, nice to meet you, Andrew."

When I got back to my office, Bob Cornell was pacing in front of it. "Where you been?"

"I just had lunch with Andrew something, stepping-stone and all that."

"Forget about him. Let's go meet some more people."

The first stop was to see Margo Alexander, the Director of Research, and, ultimately more than Bob, my boss. Margo was one of the highest-ranking women on Wall Street. "You don't have a problem working for a woman, do you?" Bob had asked before I started.

"No." I think I had a problem working for anybody, but I kept that to myself.

Margo was nice enough. She had been a retail analyst, following Sears and the Gap and those kinds of companies. She welcomed me to her department. Margo reiterated that my job was to predict the future and that I only needed to be right 51% of the time, then wished me luck. "You'll need it," I thought I heard her say as I left her office.

We looped around the department and I met an oil analyst, auto analyst, insurance analyst, beverage container analyst and bank analyst. Anything you can think of and someone was analyzing it.

As we turned a corner, the noises down the hall got louder and louder, until we finally walked out onto the trading floor. Packed into a space the size of a basketball court were hundreds of people, on the phone, running around, yelling numbers and fractions and names. Bob leaned over and said loudly, "It's quiet right now. It's only two o'clock, but it will pick up nearer to the close of the market at 4 p.m."

I met with the two head salesmen, Tom McDermott who ran the morning meeting and John Kelleher, who was the top salesman. Mind you, I still had no idea what that meant. I shook hands briefly with Robel Johnson, the listed block trader, who dealt with stocks that traded down at the New York Stock Exchange. I had heard of that. I also got a wave from half a dozen people on the "over-the-counter" desk, who traded something called NASDAQ stocks. Someday I had to figure out what all these people did.

Candy, the personnel lady, was right to ask what program I was in. Most people hired on Wall Street join right after graduating from college or business school. They get hired into a training program, where they get shifted around to various departments, such as sales, trading, arbitrage, banking, or retail, until someone likes them enough to ask them to work there. Or they complete the program and end up in an empty slot, usually sales. These programs are a great way to figure out how the system works.

No such luck for me. I was an analyst, sink or swim. Figure

it out for yourself. I spent the rest of the week organizing my paper clips and watching Jack and Steve and Jonathon Fram and even Curt Monash, the man-child, in action. They were mostly on the phone.

Bob came in on Wednesday morning. "We're going to California, first thing Monday morning. Got a problem with that?"

"Nope."

"Good. Carry-on only, no luggage."

"OK." My duffel bag would have to do.

. . .

I met Bob at the American Airlines counter at JFK airport at seven on Monday morning. We were headed to Los Angeles. Bob had the *Wall Street Journal,* the *New York Times,* and *Investors Business Daily* tucked under his arm. I had the *Sporting News.* Hey, it was baseball season—I had my priorities. That would soon change.

Luckily, Bob was a sports fan, and we chatted about baseball until Iowa went by below us. Then he dug into his papers. He didn't really read any of the articles, just focused on a page filled with earnings, shaking his head and mumbling things like "shoulda bought it," and "doesn't make sense," and "they're lying."

He finally put the paper down, looked at me, took a deep breath and launched into his story. It turned out that like me, Bob had a technical background. He, too, had studied to be an engineer. He tripped across a job as a research analyst at White Weld, a small brokerage firm. There he worked with Ben Rosen, who was a great technology analyst as computers were going mainstream with businesses in the '70s. Bob had been the top-ranked electrical equipment analyst (think General Electric) on Wall Street for ten years running, and could do it in his sleep. Bored with the grind, he had taken on the challenge of building a technology research team for Paine Webber. He told me that I had felt familiar, and that he saw a lot of him in me (without the deep booming voice, I suppose.) Wall Street, especially research, was a game, and he would teach me what he could about how to win it.

I asked a few questions, but mostly about where we were headed. Los Angeles was just a quick stop to see a company from Bob's PA (personal account), and then off to San Francisco and Silicon Valley for several analyst meetings.

. . .

After a quick lunch at an El Pollo Loco, I was about to visit my first company as an analyst. Oddly, it was a milk company, Foremost-Knudsen. Bob explained that we were about to enter a company with sales in the multi-billions, but whose stock was trading for less than $100 million in value. "We don't get a lot of visitors from Wall Street," the CEO muttered from behind a fake smile. We got a tour, a frozen yogurt tasting, and after a few hours I had stitched the whole story together.

They had billions in milk money revenue from which to generate profits, but 97% of their sales went right back out as expenses to pay farmers for the milk. They kept a whopping 3%. It was tough to make much of a profit in a low-margin business, ever. Bob owned the stock and was "long and wrong," and I think that concept hit him as we headed towards LAX for the flight to San Francisco.

Hey, my first lesson ever as an analyst, and I didn't even realize it. Revenues mean nothing. Margins and profits are everything.

. . .

That was a great warm up. That night, we drove from San Francisco airport down to Palo Alto and checked into the world famous Rickey's Hyatt Hotel. I'm pretty sure it was built before the gold rush. My room had a cold draft that felt like a wind tunnel. First thing the next morning, we were up and at 'em, and off to an analyst meeting for Apple Computer.

Arriving a few minutes early, we put our stuff down on seats in the back row of what was a rather small room. Within a few minutes, an entourage from Apple walked in and began mingling in the room. John Sculley, the ex-CEO of Pepsi whom Apple founder Steve Jobs had hired to turn around Apple, came over and talked to Bob and me. "Hi, I'm Bob Cornell and this is Andy Kessler." I think we chatted about the stock market and Apple's stock being down as well as IBM's new PC AT machines.

When the meeting started, John Sculley went over Apple's outlook, their new products and financials. I ate it all up. It came time for Q&A, and there were some silly questions about tax rates, and a softball from an analyst named Michele Preston about software. John Sculley pointed right next to me and said, "Bob, do you have a question?"

Bob shot me a smirk before asking in that deep booming voice out of his small body, "John, can you tell me about the management line-up at Apple?"

Sculley looked a little agitated, then launched into a speech about some recent problems and turf wars in the company between various groups, and how he was brought in to run the company. He ended with the bombshell "I've got to run this company as I see fit and I need to lead this company and therefore, there is no place in Apple Computer for Steve Jobs." That brought a hush to the crowd.

There were a few more questions about future product direction and licensing the Mac operating system, which John Sculley deflected to Jean-Louis Gassee, who was the head of such things at Apple. Feigning ignorance of their strategy and even the English language, the Frenchman didn't answer any questions. The meeting broke up, but management stuck around. Although I wanted to ask Monsieur Gassee about product issues, such as what chips they used, he stood in a corner, shrugging his shoulders, doing his best Inspector Clouseau imitation, repeating "I dieu nut kneeaauuww."

Sculley's "no place for Steve Jobs" line was on the front page of every paper the next morning. Bob and I went to see several other companies, as well as to an Intel analyst meeting. I met Intel's three founders, Gordon Moore, Andy Grove and Bob Noyce. I also met a cast of characters who would follow me around like a bad dream for the next decade. They were the other semiconductor analysts around the Street, my competitors: The number one analyst, Alan Rieper at Cowen, Tom Kurlak at Merrill Lynch, and Jim Barlage at Smith Barney. "These are the enemy," Bob explained to me, "these are the *Institutional Investor* All-American analysts. Destroy at all costs." They looked like a bunch of grumpy old men to me.

Intel didn't have very much good to say. Business was rolling over from the PC boom of 1984. Prices were awful.

There were a lot of new competitors in memory chips, mainly from Japan. IBM was still buying but they could turn things off quickly. These guys were brutally honest. No one got fired like Steve Jobs had the day before, so it was a let down. Intel had a director of investor relations, Jim Jarrett, whose job it was to deal with investors. He was the point man. We swapped cards, and he wished me luck. I thought I heard him whisper, "You're going to need it."

CHAPTER 2

Piranhas

I got back to New York and settled into the grind of figuring out how to be an analyst. We decided that the first company I wrote on would be one we had visited the previous week, LSI Logic. They had recently gone public and no one on the Street could quite figure out what their products, gate arrays, really did. I knew a bit about these custom chips from my days at Bell Labs, so my first task would be to explain to investors what the heck this gate array stuff was, and whether they should buy the stock.

To me, the technology was easy, and Bob beat it into my head to always stay ahead of the technology curve, since that is what investors would pay me for. But, I needed a crash course in financial reporting and Wall Street lingo. Words like consensus, top of the range, pound the table, expectations, price target, rest of the Street, reiterate and guidance didn't roll off my tongue. Not yet anyway. But soon I'd start thinking in this language.

I spent the month of August getting my first report ready. Meanwhile, I hung out with Jack Grubman and Steve Smith. I'd just sit in their office and listen to their side of phone conversations. It was still confusing.

. . .

"So, Bob, how do you figure out what a stock is worth?"

"Ah, finally, it's time for our conversation about the birds and the bees of the stock market."

"You think I can handle it?"

"We talked about profits, right?" Bob went on. "Well, the value of a stock is nothing more than the sum of all those future profits."

"That's it?"

"OK, it's a little more complicated than that. Next year's earnings are worth less than this year's earnings—you know, inflation and stuff like that make them worth not as much. They have to be discounted before they are added to the sum."

"That makes sense. How do you discount those future earnings?"

"You use a discount rate."

"Great, so there is a formula. I like this, I do math. What is the discount rate?"

"It depends," Bob said.

"Uh-oh, depends on what?" I asked.

"Future inflation, interest rates, risks."

"So is the number listed in the *Wall Street Journal*?" I asked.

"That would make it too easy."

"So what is the number?"

"No one knows. That's what makes the stock market so mysterious. Nobody knows what the future earnings of a company will be. And there is no specific discount rate. Everybody makes up their own number. So there is no real answer of how to value stocks."

"None?" I asked.

"Nope. You are just as right as the next guy. That's why you stick with fundamentals. The only thing that you can contribute to the stock market is the potential for profits from your companies. That is what investors will want to know. Usually this year's and next year's earnings are all anybody will care about. That's the duration of the market, a year and a half or two years tops. Those are the only numbers that will affect the sum of those profits."

"Usually?" I asked.

"In a recession, no one wants to look out more than one quarter, maybe even more than one month. No one will believe estimates further out. Stocks can be valued on how much cash a company has, rather than how much money they might be able to maybe, possibly make. That's no fun. On the flip side, sometimes the market has a long duration."

"What does that mean?" I asked. I was getting it, but slowly.

"A long duration means that stocks are valued based on how much people think a company will make five years from now."

"How do you know what the duration is?"

"You never know. You just feel it. Sometimes you can see it in the P/E multiple."

"The what?"

"The price to earnings multiple," Bob said. "Take the current stock price and divide it by the company's earnings and you get the P/E multiple. A high multiple usually means long duration, but not always."

"This is a weird business."

"It beats digging ditches."

. . .

On a Thursday afternoon in early August, Margo Alexander had an outing for the entire research department. It took place at her house in the Hamptons. Across the road from a secluded beach, was a beautiful spot, with gently rolling lawns, and an outstanding, award-winning flower garden. Within half an hour, we had a game of touch football going on the lawn. There were more than a few bodies tossed into the flower garden. I remember the pained look on Margo's face.

The beers were flowing and the touch football game got more intense. Some rotund guy with a cheesy moustache who I had never seen before was quarterbacking the other team. He clearly couldn't throw the ball. After doing my five-Mississippi count, I shot the line and two-hand touched him down. "What's your name?" he asked me. I thought he was going to introduce himself.

"Andy Kessler."

"OK, Andy. You're fired." Ha-ha-ha he roared.

I got back to our side of the line and asked Jack Grubman, "Jack, who is that fat asshole?"

Jack grabbed me by the neck, threw me to the ground, jumped on me and told me to shut up. "That's Ed Kerschner."

"So?"

"He's the equity strategist. He really can fire you," Jack answered.

"But I've never met the guy before."

"Just shut the fuck up, and pretend the guy is Bart Starr."

It took only two months on the job for me to learn I can get fired for upstaging a self-important Wall Street hump.

. . .

Back in the office, I learn by osmosis this whole research and analyst thing. I discover the morning meeting for the institutional equity sales force. These are the guys who call on banks, pension funds and mutual funds. Every day at 8:15 a.m. (8 a.m. on Mondays to make room for strategists, like Mr. Hot Air Kerschner), the sales force files into a conference room and hears various calls, first from traders as to what blocks they may have to move that day, and then from analysts, with changes of opinion, changes in earnings estimates or just general updates to a story.

This is the most important venue for analysts. Institutional clients do lots of business and pay the bills. Equity means stocks. There is another floor dedicated entirely to bond traders, bond salesmen and even a small number of bond analysts. A couple of microphones are scattered around the morning meeting table, as this meeting is broadcast to offices all around the country and across the world. There may be only 100 people in the room but thousands hang on your word in various brokerage offices from Sheboygan, Wisconsin, to Stockholm. You get this morning meeting right and you get the job right. It is the gateway to the outside world.

Just from watching body language, it is pretty obvious that some analysts are liked and others aren't. You can tell by how long Tom McDermott, the salesman who runs the morning meeting, gives them on the open mike. You can tell by facial expressions. And you can tell by who is taking notes versus reading the *Wall Street Journal* while an analyst is talking. I figured out quickly that maybe 5 out of 45 analysts were any good.

. . .

So what is it that makes an analyst?

Let's start with the basics. First, there are absolutely no qualifications whatsoever for an analyst job. I've always thought that a monkey could do the job, and many do. There are very few analyst training programs, and no obvious way to get a job as an analyst. Most are in the job by accident, as I certainly can attest to. Most analysts today got their job because they: 1) told an interviewer they were really smart and could prove it with a top flight B-school MBA diploma on the wall, 2) were an assistant to an existing analyst, or, in the rare case, 3) actually know something about the industry they cover, because they have industry experience. None of these paths are better than the others as each of the above can make great or lousy analysts.

Research departments at brokerage firms were initially set up to help clients pick stocks that are going to go up and avoid stocks that might drop. Pure and simple. Of course, like all good things, that mission has been bastardized over the years, so that today, even though it may be important to an investor, picking stocks is one of the least important tasks for analysts.

A guy named Barton Biggs put together the first All-Star analyst team at Morgan Stanley in the late 1970s. Ben Rosen, the hot technology analyst that Bob Cornell worked with, and who later founded Compaq, was in that group. So was Ulrich Weil, who followed big mainframes. There were also all sorts of other top analysts in various industries.

Biggs paid them astronomical salaries rumored to be in the six figures! But even then, the pace on Wall Street was slow. Remember, this was before spreadsheets, and creating income models to estimate a company's earnings meant taping together lots of graph paper and going through several #2 pencils and erasers.

Most big firms on the Street hired analysts covering all the major industries—oil, chemicals, transportation, computers, beverages, etc.—much like the S&P weightings. In the '70s, following the lead of Barton Biggs, *Institutional Investor* magazine came up with a quirky poll called the "All American Research Analyst poll." Every May, they send out polls to hundreds of buy-side firms. The sell-side is Wall Street, Morgan Stanley, Merrill Lynch, selling research and other services. The buy-side is institutions like Fidelity and JP Morgan, who buy this junk, paying with commissions. *Institutional Investor* magazine asks these institutions to rank the top three analysts in each industry sector. It's purely voluntary. Many just chuck the poll in the garbage. Others take it seriously. You never knew which firms responded.

The results of the poll are published in the October issue, by industry segment. Getting ranked in *Institutional Investor* magazine (*I.I.* for short) is the Holy Grail for analysts. Big brokerage firms love to brag that they have a highly ranked research department. Lo and behold, most research departments began hiring analysts to cover "*I.I.* slots." I filled the semiconductor slot, and the pressure began on day one to "make *I.I.*" (or die). Supply and demand being what it is, prices for good analysts, for ranked analysts I should say, keep going up.

• • •

It wasn't hard for me to figure this out. Clients vote. So let's go meet some clients. I got the report out on LSI Logic. I originally had a positive rating on the stock, not that I was really sure about the implications of that move. The week before the report was due out, the stock went up 20%, so Bob Cornell and I decided that it would be best not to go out on a limb. Instead, we rated the stock a Hold, i.e., we like it, but not enough to tell investors to load up. Unfortunately, there was a mistake in editing the report—while the rating on the front cover was changed to Hold, on the inside, in the first paragraph, it said Buy. I had the perfect hedge.

Of course, no one was going to listen to me. I was a young pup, wet behind the ears, with zero experience and zero track record. I had to explain the report at the morning meeting, initiating coverage with a Hold rating. Lots of salesmen's eyes rolled, and very few took notes. It was a yawner and not a hard-hitting call. But this was October 1985, and this report got my name out there.

• • •

The chip industry itself was doing horrible. A boom in 1983 and 1984 left the business with far too much capacity and inventory. Prices were terrible. The Japanese finally got their act together, and were cranking out memory chips, which they were selling cheap. Too cheap, in fact, and perhaps even below cost.

The first thing an analyst does or should do, Bob beat into my head, is immerse oneself in their industry. You have got to get to know every company—products, services, management, issues, quirks, gossip, *everything*. Company visits, conferences, trade show, reading industry rags, anything and everything is gathered and absorbed. From that, a reasonable judgment of the fundamentals of the industry is gained, and a coherent strategy on how to invest in the industry is developed. From that knowledge, you either recommend a few stocks to buy, or a few to avoid. So easy, right?

I would bug Bob Cornell to tell me what it takes to recommend a stock, make it a Buy, and he would just repeat again and again that it rarely had anything to do with the current stock price, but that one must focus instead on the fundamentals of the industry and each company. Was news going to get better? Were earnings estimates going to go up? Was there a sustainable recovery? The more we looked at the industry, the more we agreed that the answer to all those questions was, No Way.

• • •

With my first report out of the way, Bob took me back for a meeting with Margo Alexander, my real boss. The meeting was cordial enough, congratulations on your first report, now what are you going to do about making *I.I.*? I didn't talk much. Bob explained to Margo that the industry was going to see some better news near-term, and that orders were would get better, but that it was a false recovery, and these stocks were going to get killed. We were going to go out the next morning with a sell rating on the group. We didn't get a "good luck" when we left her office, but I did distinctly hear Margo say, "I hope you know what you're doing."

To this day, I'm not sure how Bob Cornell knew what was going to happen. Maybe he didn't. Maybe I was just a pawn and expendable, and it was worth going out with a rare Sell call, since the rest of the Street was recommending the stocks.

The next morning, a week after my yawner hold recommendation, I went back to the morning meeting to make the pitch. As it was Halloween, plenty of folks were in costume, not necessarily the serious tone I might have hoped for. A few eyes still rolled, but it could have been an important call so everyone took notes, and made the call to clients.

Now I had something to talk with investors about. I quickly built my case. Distributors were buying, not end-users. IBM had a year's worth of inventory. Prices were terrible. The Japanese were buying market share. It was all very plausible but next to impossible to prove. But hey, I had my call, and I was sticking to it.

. . .

Bob Cornell strolled into my office the next day with a funny look on his face. "I've got good news and bad news."

"Yeah?"

"Well, an important client wants to see you. Hank Hermann at Waddell and Read in Kansas City. You're going out tomorrow."

"And the bad news?"

At this point, Jack Grubman strolled in laughing his head off and chimed in "Hank is known as the Piranha. He chews up and spits out every analyst who comes into his office, the Piranha tank. You're toast."

Great, my first client meeting, and I've got to fly all the way out to Kansas City to get my ass kicked. It would be trial by fire. For the rest of the day, what seemed like the entire research department stopped by to take a look at me, before I got eaten alive the next day. Andrew Wallach even came by to remind me that this was all a stepping-stone to something else.

The salesman who covered the Midwest met me at the airport. As we drove to my execution, he made jokes about the architecture of Kansas City: Early American Warehouse. It didn't help. I started sweating.

We walked into Hank's cavernous office, in a building that was probably a warehouse. I went through my pitch, talking about a distributor-led recovery, etc. Hank sat quietly, taking some notes, and waited for me to finish. Then he calmly told me that he had liked my research report, and my Sell call. He told me that I was probably wrong, but that it took guts to say Sell and that he wanted to meet me. He then asked me a few questions about my background and that was that. I must have had an incredulous look on my face, like, "That's it?" So he got a slightly demonic look on his face and asked a few hard questions that I handled fine, and then thanked me for coming and told the salesman to make sure Hank heard everything that I had to say.

I was no longer a virgin analyst, and it didn't even hurt. It soon would.

. . .

As word got around that I could "handle" Hank Hermann, calls poured in to set me up with clients. Bob Cornell told me to take it slow, and that Hank's niceness was a fluke. He said that I should go to the outlying areas first, hone the story and then work my way to the important clients.

Steve Ally was a junior salesman out of Chicago. A recent graduate of the University of Wisconsin, where he had been a star wing on their national championship hockey team, he was a local hero. He wanted me to speak at a dinner he was holding in Appleton, Wisconsin. I had to get out a map, and still couldn't find it. I think this passed the Bob Cornell outlying area test.

I flew out to Chicago, picking up the *Wall Street Journal*, *New York Times* AND the *Sporting News*. I met Steve in the Chicago office and we left in the afternoon for Appleton. He had started recently and was trying to get off on the right skate. It was a fun trip. "Make sure you sit next to the guy from the First Bank of Neenah," he told me. "They had one of the top funds in the country last year."

I did, and learned that Neenah, Wisconsin was the home of the Neenah Foundry, a world leader in municipal castings. "What is that?" I asked.

"Uh, manhole covers."

I've checked every manhole cover I've stepped on ever since. Sure enough, they are all made in Neenah. As we left, Steve Ally told me I gave "good talk," and he thought I had a future in this business. Turns out, he did too.

. . .

Life was about to suck. Those stocks that I had a Sell call on, well, it turned out they started going up. And up. And up. It didn't matter. I stuck to my guns. I started hanging out with the traders to see if I could figure out why they were going up. Robel Johnson, the listed block trader who dealt with Motorola and Texas Instruments, told me there were more buyers than sellers. No shit? One guy was buying TI every day for the last month. There was no supply, no sellers, so the stock went up 1 to 1½ points every day.

As an analyst, you are judged right or wrong every day that the market is open. If you say Sell and the stock goes up that morning, you are wrong. Someone could have waited and sold it for more. Even worse—if you say Buy and it goes down, you are a dope. It's high stress. Don't believe anyone who says it isn't.

The morning meetings got testy. Tom McDermott, who ran the meeting, liked my call and kept having me back on. Ollie Cowan, a senior salesman who covered lots of big money New York accounts, would pepper me with questions, which were probably the same ones his accounts were blasting at him. One day he broke down and started yelling at me during the morning meeting, on the open mike. "Your stupid stocks are in a commodity business. Everyone knows that commodities are cheap when the P/E is infinite and expensive when the P/E is low. You are just plain wrong. You blew it."

I was too naïve to know if he was right about all that or not. The commodity stuff was right. An infinite P/E or price to earnings ratio means that companies are losing money, so the divisor (earnings) is zero, and any price divided by zero is infinite. So he was saying: buy commodity stocks when they are losing money, and no one cares about them. You sell them when the P/E is low, meaning they are making tons of money, the earnings are huge, and the P/E ratio is low. Some investors might think the stocks are cheap and be enticed to buy them because of the low P/E, but smart investors are selling, figuring commodities are cyclical and the earnings aren't real. But did it mean anything for me? No, it was too late to change. I stuck it out and went out on the road to see real accounts, making trips to Boston, Texas, Philly/Baltimore, San Francisco, Chicago and Minneapolis.

I was on a marketing trip to Texas, and finishing up an afternoon meeting. My client kept checking his watch, because he didn't want to miss his 5:30 dinner, with the #3 *Institutional Investor* poll coal analyst on the Street, he bragged. It turns out we were both going to the same restaurant for dinner, so I went over with him and got introduced to the #3 coal analyst, who was from Merrill Lynch. Though the analyst seemed very nice, he was a young pup like myself. I was impressed and a little depressed, because in my industry, where there were dozens of analysts, being #3 was hot. I got back to my office in New York the next day, and did a little checking. Sure enough, there were *only three* coal analysts on Wall Street.

• • •

By watching other analysts in action, I figured out there were three types of analysts. There are: 1) those who know somebody in their industry, 2) those who know their industry and 3) those who don't know anybody or anything. Lots of analysts had industry contacts—perhaps CEOs who they were buddies with or someone in the CFO's office who was feeding them information. I didn't know anybody, the best I could do was the director of investor relations, like Jim Jarrett—but everybody knew him. That was his job.

I figured I had a shot at the middle group, because I know

my industry, although a few more points higher for Texas Instruments, and that might be in doubt. It took me a long time to figure out that most of the analysts fell into the last group, and were just making it up as they went along.

But for me it was back to the morning meeting to make my pitch—distributor led recovery, IBM sitting on inventory, yada, yada.

• • •

One guy who straddled two of the groups was Jack Grubman. He had worked at AT&T and clearly knew the industry, inside and out, all the issues about deregulation, the Federal Communications Commission, competition, everything. He also had lots of old friends back at AT&T, and wasn't shy about telling anyone who would listen that he had lots of AT&T contacts. It was his edge. Not his only edge, but a big one.

Jack Grubman had this uncanny ability to predict AT&T's earnings, almost to the penny. He was either brilliant or connected. Usually about a week before AT&T reported, Jack would start whispering around the office, on the phone, and in the halls. He clearly had something. He could have gone on the next morning meeting, set his estimate at the number he was uncannily predicting, and every salesman would have made the call to every portfolio manager and trading account. The stock would have gone up or down accordingly, and the Street would move onto the next call.

But that was no fun. A better idea might be for the block trader to position the firm's capital to make money on Jack's prediction. If his estimate was for a number better than the rest of the Street's predictions, the trader might want to own the stock. However, it was probably best not to be buying the stock outright—that would be too obvious. Others would see the trades and figure it out. Instead, the desk might buy call options, sell put options, and create a synthetic long, a derivative of sorts.

No one in management complained about this setup. It was their profit and loss statement. Their bonuses were being paid out of this pool of profits. As long as no rules were being broken, it was kosher.

. . .

Jack Grubman's prescient calls on quarterly earnings were getting better. In early 1986, he was spending time analyzing a company named U.S. Telecom (UT). Jack figured in two weeks, they were going to report 25 cents per share, well over the 22 cents consensus. The stock was going to pop. I had watched this take place on the trading desk for some time, and it was time for me to play.

So I bought a bunch of options for 25 cents, the right to buy UT shares at $50 when the stock was currently at $45, so-called out of the money calls. If the stock went to $55, I would make 40 times my money. Over the next week, UT shares did go up, first to $47, then $49. My options took off, to $2, then $2½. I figured I had a down payment for a ski house in Vermont. Patience. Patience. When UT finally reported, it was 25 cents, but since seven of those cents were from a one-time gain, in reality, they reported 18 cents in operating profits. This was well *below* the 22 cents the Street expected. The stock dropped to $42 in a big hurry. My options dropped to ¹⁄₁₆th, a teenie. I ran into Jack later that day. He obviously felt like shit, having hosed a bunch of institutional investors, let alone schmucks like me. He asked if I had made money on the options. "No, I still owned them," I replied.

"Well, you could have sold them for a profit, right?" Jack asked.

"Uhhhhhhh, yup."

I was the sucker and that was last time I ever tried to game quarterly earnings. U.S. Telecom was bought by GTE in the spring of 1986, and with all his work on UT, Jack now had an entry into an old-line telephone company.

Jack was the boastful sort. In many ways, that was part of an analyst's marketing job, convincing investors that you were a hitter, a big deal, and the ax in a stock. Once you are the ax in a stock, people listen whenever you speak, and stocks go flying around. Jack got to the point where he could raise his earnings estimates on AT&T at the morning meeting and the stock would pop 1¼ within a few minutes of the opening of the market. This was intoxicating stuff. Jack's earnings calls were so good, and combined with his analysis of the telecom industry, he became the ax. Of course, you have to be right to stay the ax.

Jack's boasting had few boundaries as the pressure built to be larger than life. At one beer-soaked fest after the market close at Ben Benson's, Jack was telling a couple of traders how he had bagged three women the night before: a client, a reporter for the *Wall Street Journal* and his girlfriend when he finally got back home. "Wow," was about all the starry-eyed traders could say. The legend of Jack Grubman was growing.

I was amazed myself—until I remembered that I had been out with Jack late the night before, pounding beers and yakking it up. I got Jack alone at Ben Benson's and told him that he had been out with me the night before, not bagging three women. He looked me straight in the eyes and said, "I know. Shut up."

. . .

I had one accidental shot at glory with a quick-trade earnings call. Motorola was due to report one day at 11 a.m. They would release their revenue and earnings per share on the broad tape, on Dow Jones Newswire, but they were experimenting with putting detailed numbers and comments on the online service, Compuserve. I began checking Compuserve at 10:45 a.m., but nothing appeared. A little after 11, a detailed press release was up on Compuserve. Motorola beat revenue numbers by 5% and earned 40 cents instead of the 32 cents the Street consensus thought they were going to earn. My estimate was 35 cents, higher than anyone else, so I was ecstatic. I ran out to the trading floor and over to Robel Johnson, the block trader, screaming 40 cents. He looked at me as though I was smoking fax paper. "Nothin' on the tape, asshole."

"Forget that," I told him, "it's on Compuserve, 40 cents, revenues better by 5%."

He jumped into action, buying as many shares of Motorola as he could find, on the exchange, from other trading desks, wherever. The stock quickly popped 1¼ points.

Someone handed me the phone from Robel's turret. It was Ed Gams, the investor relations guy at Motorola. "We hear that you guys somehow got our earnings numbers. They haven't been released yet."

"Ed, it's been released. It's up on Compuserve," I pleaded.

"That was a mistake, I've taken it off Compuserve. The lawyers are slow releasing the broad tape version. Oh and Andy, you know you can't trade off that Compuserve information," Ed Gams commanded.

"Uh, Robel, Motorola just told me it was all a mistake, and we can't trade on the info," I said.

I'm pretty sure I heard Robel use the word "fuck" 14 times in one utterance directed at me. He had to unload his entire Motorola position. I only heard one part of the conversation, but it was something like, "I'm cuffed, you take 'em." All the other block traders on the street knew the 40 cents number by now as Robel had blabbed it when he was done buying. Now he had to sell his position to them. Half the street enjoyed the next 2½ point pop when Motorola finally reported 40 cents ten minutes later on the broad tape.

I wanted to be like Jack, but I think it was time for me to focus on how to predict earnings a year ahead of time, not five minutes ahead.

• • •

Jack Grubman got better and better and his legend grew. His stock calls were working. His near-term earnings estimates were working. He was becoming a star. One of the tools of the analyst game is the conference call. These were new to Wall Street. You announced a conference call and 100 investors from around the country called a number and were connected to an analyst, who went through their pitch and took questions. While they are commonplace today, they were fairly new in the '80s and getting cheaper every day. On one of these calls, Jack got a question about alternate pay phone companies. These were independent operators who had found a quirk in the law and were putting in phone booths and pay phones around cities, charging exorbitant rates.

Jack dutifully answered the question. "They are a bunch of crooks stealing from legitimate operators."

Now it turns out that this was somewhat true. But the person who asked the question, and obviously owned stock in alternate pay phone companies, went berserk. He yelled at Jack, hung up from the conference call, called up and yelled at Margo, and then called Paine Webber CEO Don Marron and probably yelled at his secretary. Margo came down and asked Jack, "Are they crooks?"

Jack answered, "Yeah, they lie and cheat and put in these boxes that look the same and . . ."

"Don't worry about it," Margo cut him off.

An analyst who is making money for the firm is to be protected, no matter what. Clients can yell all they want. I learned a great Wall Street adage that day. Jack asked me, "What's the difference between the buy-side and the sell-side? On the buy-side, you can say asshole *before* you hang up the phone."

Jack made amends by meeting with the investor, hearing the story, and visiting the alternate pay phone company. They were sleazebags, but Jack was intrigued that you can make money "stealing" AT&T's business. From then on, I noticed a new Jack Grubman—one who was no longer comfortable with the mothership AT&T and looking for those companies that could beat up on Ma Bell. This would eventually lead him into the influence of Bernie Ebbers who would soon be amassing the house of cards to be known as Worldcom.

That U.S. Telecom stock, which I tried to game short-term

and blew it? I should have just bought the stock. GTE ended up buying UT for its Sprint long-distance business. A few years later, GTE bought another company, Contel. They paid $6.6 billion, which was no tiny sum in 1991. This made GTE the largest local phone company, and the second largest cellular operator. GTE CEO Charles Lee called Don Marron and told him Paine Webber could be the investment banker representing GTE on the deal. This was big—Paine Webber's last big client was Greyhound bus. Paine Webber could have the business, Lee went on, but only if Jack Grubman got paid for bringing in the deal, since the merger was Jack's idea in the first place. I don't think Don Marron thought about it for very long. It meant $10 million plus or minus in fees, which was pure profit. Jack got paid all right. In addition, Don Marron paid to have Jack's house at the Hamptons redecorated. Jack was finally big time.

• • •

Don Marron was not the most beloved CEO on Wall Street. He had merged his way into Paine Webber through a small firm he controlled named Mitchell Hutchins. He also made it big personally when he sold an economic data business, DRI or Data Resource Inc. to McGraw-Hill. No problem so far. What irked everyone at Paine Webber were his outsized pay packages, huge bonuses, option plans, and restricted stock. Clearly, the guy knew no shame when it came to paying himself. That didn't bother me in the least. What bothered me (and plenty of others at the company) is that he spent much of his fortune on art. Contemporary art. Big, bad, ugly art.

Jay McInerney's book, *Bright Lights, Big City,* was out at

the time and in it, there was a great line that said, "Taste is just a matter of taste." OK, Marron could have his own taste (I had an Ansel Adams poster in my bathroom at home.) But the guy hung up his art in the hallways and reception areas of Paine Webber; every spare wall was covered. The one that everyone in research had to see daily as they passed through reception was this hallucination of purplish upside-down people flying towards what looked like, if you squinted really hard, Caesar's Palace, the casino in Las Vegas. Perhaps this was a subliminal message by Marron that we all worked in some surreal casino.

CHAPTER 3

You're in the Entertainment Business

OK, I can't blame the bad art for what is turning out to be a bad sell call on Intel, Motorola and Texas Instruments. The sales force is snickering. Clients are cordial when I visit them, but since most own these stocks, they are glad I am wrong. Lots of analysts would pass in the hall and say, "Hey, how's that sell call going?"

Turns out not all that badly. The stocks are going up, but not that much. They are not going up any more than the market is, and it's a bull market. For every 100 points on the Dow, a reasonably common occurrence then, Margo buys a case of champagne for the department. But chips stocks are just keeping pace. Some investors start to worry that something may be wrong. My competitors keep pounding the table saying, "Buy more."

I keep marketing. It's not like I had much better to do.

And I finally figure out what this Ed Kerschner guy does. He had a super proprietary system he called the Kerschner-Pradilla model. Pradilla had been his economics professor way back when. They send out a survey to the buy-side, asking them for their earnings estimates on the 500 largest stocks. The KP model figures that the buy-side's earnings estimates are the consensus; therefore the model knows exactly what expectations really are. Then they sort the stocks by growth and P/E ratio and rank them in order of attractiveness. It's a neat concept. Too bad it didn't work for shit.

Kerschner would show up on the morning meeting every Monday, and go through his model, naming stocks that moved up or down his ranking. I think most of the sales force figured out that the model was bogus, and politely nodded without taking notes. Art Cashin, who ran the trading floor operations for Paine Webber, usually followed Kerschner. Art told fun tales of Wall Street and stock trading, in two minutes or less, that were almost always on the mark. More people listened to Cashin than to Kerschner.

Someone, probably Kerschner, had the brilliant idea that each industry group should meet with Kerschner so he could tell them that they were wrong and that his model was right. Jack and I showed up on time for ours, at 2 o'clock one day. Some research assistants were there, but Steve and the rest of the analysts were traveling. Kerschner puffed in around 2:20, with a look like we were wasting his time. He looked around, said to everybody present, "Nobody is here," turned around and left.

"He's still a fat asshole, Jack," I said.

"Yeah, more of a Sonny Jurgenson or Billy Kilmer," Jack replied, referring to two bloated quarterbacks, "than a Bart Starr."

The group started traveling together; we would show up in a city and invite investors to a Technology Group lunch. It was great for me. Steve Smith was the top computer analyst, Curt Monash was a respected software analyst (until clients met him), and Jack Grubman was getting more famous by the day.

• • •

Jack Grubman was a big time boxing fan. On dull afternoons, he would sit in my office and tell me about his days as a Golden Glove boxer in Philadelphia. He was pretty good, he told me, until the day he got his nose flattened with a knockout punch, and ended up out cold on the mat. He got up and quickly decided that what he really wanted to be was a security analyst, not a prizefighter.

We talked a lot about boxing. Ali, Frazier, Foreman, Holmes, Hagler, Hearns, Sugar Ray Leonard, Tyson, Biggs. We made a pact to go see some prizefights, wherever they were.

One of the greatest stops for analysts, Alliance Management, was in Minneapolis. When you arrived, you were ushered into a conference room, where a very large sign read, "State your conclusions upfront, we'll decide what to talk about."

Al Harrison, who ran the big portfolio for Alliance, would come in, sit down, hear your conclusions, and then fire questions at you, each one better than the next, to get out what you really thought about your industry and various companies. It was exhausting, but when you were done, you either held your own, or needed to go back and rework your thinking. To me, these were the smartest guys in the business, and they had the performance to back it.

On one trip to Minneapolis, Jack, Steve and I were out to dinner with a client, and then retired to some bar to stay out of an ice storm. At about two in the morning, we rolled back into the hotel. In the elevator, some guy started jabbering to us, calling us "suits" and going on about ice fishing or something. The guy got off a floor or two before ours, and I had to hold back Jack, who began to lunge after the guy. Jack turned to me and asked, "Should I have, like, decked that guy, or what?"

"Take it easy, Jack."

• • •

Very few people wanted to hear a Sell call, so I decided I needed to recommend something. I had been doing work on a small company in southern California, named Silicon Systems. I visited a couple of times, met with the CEO, Carm Santoro, and went through the story. They had some hot chip for dialing. I worked up a report and, on December 30, when things were pretty empty and sleepy at Paine Webber, I got time on the morning call and recommended the stock.

It went up that day three-eighths of a point to $9. A portfolio manager at T. Rowe Price in Baltimore who owned 7% of the company called me and thanked me for moving the stock three-eighths. I think he was pulling my chain.

A week later, Silicon Systems put out a press release saying they were going to miss their December quarter earnings. They had pulled in revenue to make the last quarter, couldn't make up the shortfall and went from a modest profit to a huge loss. The stock dropped from $9 to $6.

Wasn't I the biggest moron in history? I had a breakfast meeting that morning, so I sent a note to Tom McDermott explaining the shortfall, and reiterating my Buy. Tom read the note, paused, and then said, "Welcome to Wall Street, Andy."

Everybody at the meeting, as well as everyone listening in

at branches around the world, got a good laugh, except for me.

I was just under six months into my new so-called career as a Wall Street analyst, and so far I'd been dead wrong. I had a little more than six months to be dead right to reach that "get it right 51% of the time" thing.

Then a funny thing happened. In the spring of 1986, chip stocks rolled over. They did not just stop going up, they started going down. And down. And down. Every damn one of them. No one could figure out why. Of course, this got all my competitors pounding the table at their own morning meetings, reiterating their Buys. Big mistake.

Ed Kerschner presented at the next Monday morning meeting. AMD, one of the stocks I had a Sell on, was his top pick. The model doesn't lie. The stock had gone down before the earnings and growth inputs to the model had changed, so, at $35, it was attractive in the model. I felt lots of eyes on me that morning. I didn't say a thing. Within six weeks, AMD would be $16. Ouch. I learned a new lesson. Never trust models. They are garbage.

• • •

Though I was starting to be right on my stocks, I was increasingly convinced it didn't matter. I looked around the research department, and noticed that very few analysts were right about anything. The job of an analyst has more to do with impressing everyone they come in contact with, and less to do with stocks. At one meeting I had with Fred Kittler, a portfolio manager at JP Morgan, he paused, looked me in the eye and said, "You realize, you are not in the analysis business, you are in the entertainment business."

Damn, that hurt, but he was right.

It was that stupid *I.I.* poll. Get ranked. Get votes from the buy-side. There is only a weak correlation between being right on stocks and getting votes, and there is no correlation between providing great investment advice and generating more business for your firm. Even if there were, some salesman or trader would get credit. As an analyst, your only hope for recognition is to be ranked by *I.I.*, and to show up in a silly magazine every October. Actually, the October timing is fortuitous since Wall Street bonuses are calculated in December. You might as well peak in the fall. No use doing good stuff in February, because no one will remember!

The only reasonable correlation is between *I.I.* ranking and investment banking business. The way to land competitive and highly sought-after IPOs and other financings is for investment bankers to brag and drag in their *I.I.* ranked analyst. They promise the analyst will cover the company, write great reports, recommend the stocks if warranted (it is always warranted), and have the analyst's great name attached to the stock, which must be worth at least an extra 5–10 multiple points. Bankers will promise anything to get a piece of business! Since IPOs pay 7%, and trading stocks in 1986 pays 5–10 cents per share, an analyst has increasingly become more important as deal bait, rather than for generating trading flow. Still, the trick is to get ranked.

• • •

How do analysts get ranked? Bob Cornell laid it out for Jack and me and the rest of the group. It's Marketing 101, in order: phone calls, visits, reports, PR and, oh yeah, being right!

Almost every research director on Wall Street institutes a 100 phone calls a month program. Affectionately known as "dialing for dollars," analysts are *required* to call the top 100 institutional accounts once a month. This is tracked, put into a database, and correlated against results. Bizarre, but all of these highly paid, highly qualified industry experts spend over half of their time calling places like The First Bank of Neenah monthly because someone once heard that they vote in the *I.I.* poll. Even more strangely, this "100 calls a month" program actually works. I always hated the calls, as it took time away from figuring out what was going on in the industry.

The next marketing requirement was visiting every account once a year, sometimes twice. That meant going into someone's office, rolling up your sleeves and having 45 minutes to convince them you knew what you are talking about. Sometimes you were met with a stone face, and sometimes, as with Al Harrison at Alliance, you got great dialogue. I also had a client in Texas literally fall asleep. The guy was 82, I later found out. I kept talking.

I've had clients yell at me and throw me out of their office. I always preferred the growling client. They kept you on your toes, and made you do some homework before you met them. There were eight to ten one-on-one meetings a day, plus a group breakfast, group lunch and yes, a group dinner, plus travel, all the time saying the same tired old story. With little time to figure out anything new to say, the trick to longevity is picking a theme that can last a decade and finding new little spins on it.

Between phone calls and visits, the next consuming task is writing reports. Because these research reports are mailed out in the thousands to clients and to companies, they have to look good. Perhaps they even have to have substance, but certainly they must look good. These reports tended to get piled high in buy-side mailrooms and never read (something I never realized), but were still necessary to create the proper paper trail to prove one's worth to clients in order to stimulate their all-powerful vote. One client had a stack of research reports in his office pile six feet high. He pointed out that it was close to his garbage can so that when the stack got too high, the reports would fall naturally into the garbage, "where they belong." He also dialed his voice mail, which he empties every day, so I could listen. At 11 a.m., it was no longer accepting messages—it was over the 100-message limit.

. . .

How do analysts cut through the clutter? With all of the salesmen's calls, analysts' calls, and piles of research, it is hard to stand out.

The answer is the press. Getting in the press helps the cause—actually it is worth a whole day of phone calls reminding clients who you are and how valuable you are to their jobs. When the reporter for the *Wall Street Journal's* "Heard on the Street" column calls, you've got to give her something good or you won't get quoted. Of course, sometimes, analysts get quoted on something new and insightful, which pisses off clients who hear about it for the first time in the paper. More than once I have been scolded, "Why should we pay your firm a million bucks a year in commissions when I can get your research for 75 cents in the Journal?" Oops. Still, it got the votes.

Votes get you paid. Well, not directly. Votes got another firm that wants to rank higher in aggregate in the *I.I.* polls interested in buying your vote. Analyst pay is market-priced. One of the only ways to get a raise is to get a competing job offer.

It was a funny system. Strange, even, because every analyst I've ever talked to admitted that they were overpaid (in private anyway). The conversation usually went like this. "No way we deserve what we are getting paid, from, you know, society's point of view. Having said that, I just want it to be known around here that I am way underpaid." No wonder analysts are so wacky.

. . .

For analysts, the focus is on institutions. They pay the commissions. They are the clients. But here I was at Paine Webber, one of the bigger retail firms, derogatorily known as a "wire house" from the old days of brokers telegraphing or wiring in their stock orders from individuals. I got lots of calls from annoying brokers. What did I think of this, what did I think of that? Why is Intel going down today? Tom McDermott told me, "Ignore the gold chain and double-knit crowd."

"Who is that?" I asked.

"Retail. Individuals. They will only get you in trouble. They're not sophisticated enough to handle investing. Let me put it another way. They are stupid, the dumb money that always gets beat up. Retail waits until a stock works, and only then buys it. Then they complain when it goes down. Or they won't sell something that works, convinced it's going higher, and then blame you when it blows up. Stay away from all of them. Use garlic and crosses if you have to."

This turned out to be some of the best advice I received in my years on Wall Street.

CHAPTER 4

Bull Run

One of the fun things about working on Wall Street during a bull market is the amount of money that is thrown around. Paine Webber had this great tradition. The equity salesmen would go out to dinner at Ben Benson's or some other expensive restaurant, like the four-star French seafood place down the block, Le Bernadin. The group would run up as big a tab as possible—I'm talking thousands—and make the youngest person there put it on their credit card and figure out how to expense it. Say, 120 cab rides or 30 fictitious lunches. For some reason, analysts never counted when it came to being the youngest, so I would go out of my way to find out about these dinners and get myself invited. We'd talk about business. OK, no we didn't, we gossiped about who was any good. I learned some real dirt about how worthless many analysts were.

The day after one of these dinners, I began a marketing trip through the south, to Virginia, North Carolina, Atlanta and Orlando in 36 hours. Analysts affectionately called the trip

"Sherman's March to the Sea." I hosted a lunch on top of the tallest building in Charlotte. There were ten clients present, plus myself. Most ordered drinks and appetizers, a main course, and dessert. I was cringing, thinking about the $4000 tab the night before for a similar-sized group, and what I was going to get stuck with. The bill came to $86. I paid cash. The waitress wanted to know why I was laughing.

• • •

Margo Alexander had an offsite meeting every year for analysts only. It took place at Arrowwood, a somewhat dumpy conference center in scenic Rye Brook, New York. It had a blowout dinner the night before followed by a full day of meetings, seminars and other feel-good nonsense. The first year I attended, Jack Grubman and I decided that the tech team should host a poker game after dinner. Along with Steve Smith, we invited a select few fun analysts. We ran out of people to invite pretty quickly. I scouted around and found a small conference room that was set up for some accountants the next morning.

We immediately called room service and a woman showed up to take our order. Before anyone could say anything, I ordered 12 Heinekens and 45 Kamikazes. I had to explain to the woman that a Kamikaze was a toxic elixir of vodka, triple sec and limejuice. Everyone nodded. That would do for now. "Oh, and peanuts. Yeah, and barbecue potato chips if you have them. Oh, and do you have cards and poker chips?"

The poker lasted much too late. Forty-five Kamikazes turned out not to be enough, and several more rounds were ordered. The room got progressively more disheveled. Lights were smashed, tables turned over, curtains removed, and the carpet was not its original color. The place was seriously trashed. Keith Moon and the Who couldn't have done it better. Hey, it was a bull market. We were rock stars.

After about two hours sleep, the next morning came too fast. I slipped into a seat at the back of the room and tried to focus on the presentations. Ed Kerschner was running around demanding to know who trashed the Price Waterhouse conference rooms, because they were going to pay. Nobody fessed up. I probably didn't realize how close I was to being fired, this time for real. No matter. I was about to go from the biggest dumb shit on the planet to the smartest guy in the world.

• • •

By April 1986, orders from IBM for chips disappeared. They really were sitting on tons of inventory and stocks had rolled over when people in the industry (but not on Wall Street) realized this. I had been predicting it, but it wasn't until it was announced that industry orders were heading down, and the stocks in the group had already halved, that the truth became clear. Someone in industry always knows before the Street. As an analyst, you have to be ahead of these moves, and predict, not report, them.

One by one, my competitors began downgrading the stocks in the group. Tom McDermott had me on twice a week to reiterate that we had a sell for the last (painful) six months. The stocks didn't just go down, they went to five-year lows, lower than the bottom of the bear market of 1982. It was a disaster. Not for me, but for anyone who owned these things. Clients began returning my calls and asked when the stocks would bottom, and what would make them go back up? I had no clue.

But I was enjoying being right for a change. I got calls from the *Wall Street Journal, Barron's, Fortune,* and from *Business Week.* I made sure not to say anything new, not that there was much new to say, except to explain the downturn. Ollie Cowan, the salesman who had told me I was dead wrong, stopped by and congratulated me on the call. I got more respect from the rest of the sales force. Jack Grubman came into my office and said, "You're fucking hot. Don't blow it."

• • •

1986 was a strange year. Things were going right. It wasn't the rip-roaring bull market that would show up in 1987, but things were hopping. All of New York caught stock fever. Still, it wasn't so easy. No matter how good a call you had, most of the stocks that were working were takeovers, leveraged buyouts and companies put in play by Drexel Burnham and its junk bond king, Michael Milken. I heard about these guys every day. The rest of the street was in awe. One day, I was walking down 5th Avenue with one of our salesmen and his client, when the salesman asked why the account didn't do more business with us. The client was brutally honest. "Who needs you? I get a call every morning from my Drexel salesman who tells me the next set of companies that are going to be in play. I buy them and my portfolio soars. I don't need anyone else."

These were pretty sobering thoughts. The more you knew about Wall Street, the more confusing it was. But it was booming. People were streaming into the business, chasing the bull market. There are a number of entry exams that people on Wall Street have to take. The most basic is a Series 7 test to become a registered representative. When you passed the four or five hour test for a Series 7, you could recommend stocks and take orders from clients. Someone found out that the Securities and Exchange Commission (SEC) was changing the test questions, and the last test using the old questions was a month away. I didn't even know about the test, until it was decided that everyone in research needed to pass Series 7 and should take the next test. The reason the changing of the test questions was so important is that the Series 7 test is filled with trivia questions, like this:

The holding period for Rule 144A shares is

a) 364 days
b) 365 days
c) 366 days
d) Forever
e) All of the above

There were special classes set up to "study" for the test, but all you really needed to do was get the old questions and cram the night before the test, which was on a Saturday.

Everybody on Wall Street had heard about the questions changing, so that Saturday—in May I think—downtown Manhattan was packed with people taking the Series 7 test. All federal and state buildings, as well as schools, were commandeered as test-taking rooms for the two sessions from 10–12 a.m. and then 1–3 p.m.

Steve, Jack, and I ended up in the same room. The questions were the same ones I had studied the night before, but in a different order. I finished at about 11:15, turned in my papers and whispered to both Steve and Jack as I walked out,

"You two must be serious dumb shits. Hurry your asses up, I'm hungry."

Within 30 seconds, Jack popped up and ran out of the room to catch up with me. It was a competitive thing, I guess. Steve was not far behind. The three of us went to the Bridge Café, under the Brooklyn Bridge—Mayor Ed Koch's favorite restaurant in New York. Over one, then two, and then three pitchers of beer before heading back for the second half of the test, we soaked in the sunshine and laughed about how stupid you had to be to fail a test like this. (I wouldn't find out for another 15 years that Jack actually failed the test that day. Oops.)

It didn't take that much to work on Wall Street. A few trivia questions and you were in. You would think with all the money involved, there would be a somewhat higher bar.

• • •

Wall Street is a morning business and there were a lot of dull afternoons. Having only so many boxing stories to share with Jack, I used to head out to the trading floor and pull up a chair next to the block trader, Robel Johnson. He would often explain how things worked, but mainly he would put on a show, for me and for every other trader.

One afternoon in 1986, Robel pointed to his screen that showed he owned 70,000 shares of Sperry Univac, one of the dying computer companies that IBM was killing in the marketplace. As the afternoon wore on, Sperry kept inching up, up 1, up 1½, up 2, up 1¾, up 2½. Something was going on. Robel was enjoying himself. At 3:59 p.m., right before the close, he sold 10,000 shares of Sperry at the market. That was that. It hit 4:00 p.m. and the trading day was over. There was always some limited trading until 4:30 p.m. at the Pacific Exchange, so I hung out for a little bit longer. At 4:05 p.m., traders started yelling over to Robel, "Sperry on the tape."

Robel let out the loudest scream I had heard on the trading floor. Burroughs Corporation was merging with Sperry, up 12 bucks from the previous close. "Shiiitttttt. Goddaaaaaaaaam-mit. I sold a piece at the close." He also had a big smile on his face, as did most traders who looked over. The block sale at the close was a bit of a charade by someone who knew what was coming, or such was the impression I got.

Another day, Robel was chatting on the phone when he got this sheepish look on his face. I had never seen this from Robel before. "OK, I understand, will do, consider it done, right away."

He hung up the phone, turned to me with an ashen face and said, "That was Ivan Boesky, and he's selling everything."

"Who?" I asked.

"Never mind."

Three days later the front pages of the papers were splashed with the news of the arrest of Ivan Boesky for insider trading and for paying for information with briefcases filled with $100 bills. Robel left the firm and went to Kidder Peabody downtown. I never found out why. He was replaced by a much younger and much more serious guy, Mike Lock-wood.

• • •

On another slow afternoon I was sitting in my office, probably dozing off, when I heard this loud racket coming from the next office, which was Jack Grubman's. I heard rapid-fire "Goddamn, sonofa, mother, shit, goddamn . . ." and then an explosion. I ran over and saw Jack's phone in about 50 pieces scattered through the office. He had emptied his bookshelf onto the floor and was starting to break everything he had on or near his desk. It turned out that AT&T had reported 26 cents instead of the 28 cents that Jack had been whispering.

He ran to my office and took over my phone. "You don't know who you've messed with. This isn't some low stakes penny ante game. Our desk is long and wrong to the tune of millions." Turned out the trading desk and Mike Lockwood had unwound the position when AT&T's stock had worked too early, and even though AT&T's stock got hit after the bad numbers came out, it wouldn't cost the trading desk that much. To this day, I'm not sure if I had ever seen anyone that mad with so many veins popping out of their head.

• • •

By July 1986, the carnage in chip stocks seemed to be over. So, without thinking too much about it, I did the only thing that made sense. I flipped and went to a Buy call on Intel and Motorola. Of course, this pissed everybody off. "You should milk the Sell call while it's working," a few salesmen told me.

That didn't help me. I scrambled all evening to come up with something interesting to say. I focused on Intel and Motorola, postulating that they could sell microprocessor chips for $300 that cost them $25 to make, value pricing at its best. I went out the next morning with my call. The stock market dropped 50 points that day, and both Intel and Motorola dropped another buck. So much for an impact call.

And dammit, I was right back where I had started. Everybody hated these names now. I had finally been right and now I had to go and slog around to all these clients, and tell them to buy these names after I had just said Sell. I was always swimming against the current. I thought this was supposed to be fun.

A week after my Sell to Buy flip, Intel announced earnings. Well, not exactly earnings. They reported a $100 million loss, the biggest in their history. They were closing plants, getting out of memory chip production, and laying off thousands. Not the stuff to instill confidence in my Buy call. The stock, which I had recommended at $17½ and had nicely popped to $18¾ since my recommendation a week before, was now at $16 and change and quickly heading lower.

I ran out to the trading desk. A guy named Joey Palma was the over-the-counter trader responsible for Intel. Joey was a nice, mellow, middle-aged guy, who I figured had worked on Wall Street forever. Before I could ask him what he thought was happening to the stock, I noticed he had half a dozen needles stuck in his ear. Some weird acupuncture thing to calm his nerves? Who knows? I was too spooked to ask.

"Nice loss," he said.

"Yeah, lots of one-time stuff for closings and layoffs and . . ."

"Whatever, this puppy's going down."

I must have had a sad puppy look on my face, so Joey decided to cheer me up.

"Andy, don't worry about it. I've been trading Intel for years and years and whenever it goes down, it always bottoms around here, $15 or $16 bucks. It's almost there."

I felt relieved, like a ton of bricks had been lifted off my back. This is it, this is the bottom, the big loss won't hurt me.

My call is going to work after all. About halfway back to my office, around the Caesar's Palace hallucination painting, something hit me like a two-by-four piece of lumber to the head. Intel's stock has split a bunch of times over the years. There were two-for-one splits, three-for-two splits. Joey's comforting words were completely worthless. There was no mathematical reason for the stock to bottom here or anywhere else. It was just the psychology of the market that remembered Intel bottoming at $15 or $16. This thing could get whacked further.

When I got back to my office, I felt like shit. Jack stuck his head in my office and offered even more comforting words, "You're fucked."

So be it. That was the job. Back on the road. A hundred calls a month. I was finally right and now it was time to start all over again. Market/entertain away. Intel did bottom, at $15¾. It would be $60 within nine months. The lesson about why companies split their stock is a great one.

• • •

The legend of Jack Grubman kept growing. Some of this was his doing—spreading stories about himself was his marketing task. Some of it was what he knew. He nailed each quarter. He had contacts at both AT&T and in the industry to draw from. Peter Lynch, who ran Fidelity's Magellan fund, the largest fund in existence, called Jack up to Boston so he could teach Lynch about cellular. He had dinners. He had conference calls. We often traveled together and continued marketing as a tech team.

Jack and I started going to boxing matches. Sometimes they were small fights in New York. Sometimes they would be closed circuit TV viewings of prizefights from Las Vegas. On those occasions, Jack would host a dinner beforehand. Present at one such dinner were some interesting guests: Ed Greenberg, who was Morgan Stanley's telecom analyst and Jack's competitor, Cal Sims, a reporter for the *New York Times,* and a couple of clients. And me. Jack worked every angle. He wanted to give a right hook to our slow waiter. I calmed him down.

• • •

I was looking for any excuse to say something good about my industry, after bad mouthing it for so long. In between marketing to clients, I traveled to see every company, met every management team, and went to every trade show. I lived and breathed it all. I traveled three or four days every week. The 6 p.m. American Airlines flight from JFK to SFO was my shuttle. The flight attendants got to recognize me.

Then all of a sudden, on January 2, 1987, the market jolted upward. Intel and Motorola's stocks took off as well, and never looked back. Orders in the industry got better in January, giving a fundamental support to these stocks. July 1986 really was a bottom.

I wrote a report on Intel, "The New Alchemists: How Intel is Turning Silicon into Gold." They got rid of AMD as an alternate source on their microprocessors and sure enough, they could sell for $300 what cost them $25 to make. Margo Alexander passed me in the hall and told me she had good news and bad news for me.

"My husband, Bob, read your Intel report and really liked it."

"And the bad news?" I asked.

"He bought the stock. It better not go down."

On Super Bowl weekend, I flew to Boise, Idaho, to visit

with Micron Technology. Micron was a bit player in the industry. (Bad pun.) They just made memory chips. When things are good they make a killing. When things are bad, they get killed. At this point, they were on their deathbed. I cared about Micron only peripherally. I wanted to go skiing at Sun Valley.

I flew out Saturday night. Since Sun Valley had no snow, the CFO of Micron sent me to another place, and I drove through the night to get there. What I first thought was a tiny hill with one lift turned out to be a massive mountain dotted with trees and light powder up to my armpits. That night, with a smile on my face, I went to a dinner at Micron CEO Joe Parkinson's house, which was enormous. I spent most of the night next to this nice old man, JR Simplot. A potato magnate, he sold almost every french fry which McDonalds bought. The French Fry king was also a large investor in Micron. I asked him where he lived, and he pointed to a hill across the valley. "See that flag-pole?"

"You mean the one at the entrance to a state park?" I asked.

"Yeah, well, that ain't no park, that's my place."

Micron was in critical care. Joe whined to me about the Japanese dumping chips. He claimed they were selling them below cost to put him out of business. Intel already dropped out, but Micron couldn't, because it was all they did. The stock was $2 and heading to zero. I was the only one from Wall Street at the annual meeting the next morning. And I was really just there to ski!

Joe started telling me about an effort to "hit them where it hurts." D.C. was on his side. Something was going to happen. It was all very cryptic, like a bad spy movie.

At the airport, I watched the Giants win the Super Bowl

and flew home that night. I should have gone into the morning meeting and recommended the stock. Or bought it for my personal account, my PA. Nope. Didn't do either. But I did start sniffing around. I was intrigued by the "D.C. was on his side" comment.

I called the patriotic-sounding American Electronics Association to see what they knew. They pointed me to a lobbyist in Washington who pointed me to a lawyer. He started telling me about what Senator Pete Wilson was trying to do, and said that the Commerce Department was ready to act. "This thing goes way up."

I called them all and started piecing together a story. Indeed, something was going to happen. The Japanese were going to get slapped. Dumping duties were to be imposed, meaning there would be huge fines in retaliation for the Japanese selling below costs, and a potential trade war.

I tried all this out on Tom McDermott, to make my case on the morning call. "Are the stocks going up or down?" he asked.

"I'm not sure." I said.

"Come back when you are."

It wasn't an investor call, but it was big.

So I called Calvin Sims, the *New York Times* reporter I'd sat next to at Jack's boxing dinner. He took the call and heard me out, although he barely remembered me. "Something big is going to happen," I said.

"What do you think?"

"Sanctions, dumping duties."

"What the heck is that?" he asked.

I walked him through everything I knew. He thanked me and hung up and that was that. For about three weeks.

All of a sudden, the story started getting hot. Someone in the Reagan administration was floating a trial balloon with the press about sanctions against the Japanese. The Washington bureau *New York Times* didn't know anything about it and checked with New York to see if anyone there did. Calvin Sims said he knew a few things. He had called around and no one knew anything and then he remembered our conversation and called me up on a Friday afternoon. What did I know, what could happen, why was this happening? We went through it.

Over the weekend, I checked the *Times* for any stories. Nothing. Monday morning, however, there was a big fat, above-the-fold, front-page story in the *New York Times* by Calvin Sims about potential dumping duties. He quoted me a number of times about why the administration was going to do it and what it meant for the industry. It was a nice way to start the week.

At about ten in the morning, I was sitting in my office shooting the shit with Jack Grubman. Our secretary came in and said that a producer from ABC's *Nightline* show was on the phone. Did I want to speak with her?

"Sure, why not?"

"I know you must be very busy this morning," she told me as I cupped the phone and asked Jack if I was very busy this morning. He nodded yes. "But could we get a crew over to tape some comments by you about these dumping duties?"

There wasn't much to say, because they hadn't announced anything yet—it was still a trial balloon. I gave ABC about 15 minutes of babble, but the only line I thought of ahead of time was, "We can't be wimps in the world of trade."

Sims called that day for a follow-up story he was doing, and then confided to me that he was over his head. He was a telecom reporter and lots of people at the *New York Times* were itching to run with this story. But it was his, because I had called him, because he was at a boxing dinner. He told me he was getting thrown off the story after the next day's piece. He quoted me again, but by Thursday and Friday, the entire front section was filled with stories about trade wars and sanctions, quoting various officials in the U.S. and in Japan. I watched *Nightline* all week. Nothing.

On Friday, after the market close, there was a two-sentence press release outlining specific dumping duties on Japanese products by the Reagan Administration. The story exploded everywhere. I was on *Nightline* that night. They used the "can't be wimps in the world of trade" line. Hmmm. It was time for me to think up more sound bites and other ways to use the press to my advantage.

. . .

Downstairs, in the tunnels of Rockefeller Center, there was some wasted storage space that was taken over by a cable channel named FNN (Financial News Network). They had a camera or two, some talking heads caked in makeup, and probably 50 viewers. They would often call at 3:45 p.m. to ask me to come downstairs and appear at 4:15 for a wrap up of the day's trading in tech stocks. It was almost on my way home, so why not. I was on fairly often. Of course, this eventually turned into the CNBC of today. I should have asked to be a regular guest.

When you are an analyst, nothing happens for long stretches; but when something does happen, you need a comment in less than two minutes. There were calls in the morning, then long boring days until companies report earnings after 4 p.m. Ninety percent of life as an analyst is showing up. Our offices had a door and a narrow window so people could look in when the door was closed. In the afternoons, I used to close my door, prop my feet up, open a trade magazine, and try to fall asleep for a while. The problem was that Paine Webber had these goofy motion sensors that detected motion and turned the lights on when you entered your office. They would also turn the lights off after about ten minutes without any movements. Sure enough, I'd fall asleep and the lights would turn off. Jack would walk by, see me sitting there with the lights off, open the door, which made the lights turn on. Then he'd make fun of me for taking a nap.

One of the young assistants at Paine Webber, Henry D'Auria had a bigger problem. I came in one day at 8 a.m. and he looked like shit. Either he was out all night drinking or had slept in his clothes, or both. He told me he was about to be made a full-fledged analyst following fast food restaurants (I could have introduced him to the French Fry king!) He had pulled an all-nighter getting his presentation ready for the morning meeting, which was to take place in fifteen minutes. When he finished working on it at 4 a.m., he figured he could get four hours sleep on the floor of his office. His problem was that he tossed and turned in his sleep, and the damn lights kept coming on and waking him up. So he crawled further and further under his desk until he was out of reach of the all-seeing motion detectors. He figured that happened around 7:30 a.m. He woke up after 30 minutes sleep with his face smashed against the back of the desk.

. . .

One evening a few weeks later, Jack Grubman and I and a research coordinator at Paine Webber named Ray Ozyjowski who, for obvious reasons everyone called Ozzie, were at a downtown dinner with a client. When the meal ended, we all decided to share a cab uptown. Ozzie, who was in the front seat, turned around and announced that he was psyched to go to Atlantic City. He asked the cab driver how much he would charge to take us to AC, a drive of more than three hours. The cabbie just shook his head and mumbled, "No way." Ozzie caught my eye. "You in? We'll pull up to the next ATM and hit the limit on our cards and head to AC."

"I'm in," I said laughing.

Jack played right along. "I'm in, I can get us a room, and we can gamble at the Sands."

The client was silent, scared shitless that we were going to make him come along. Ozzie kept working the driver. "C'mon, imagine how excited your old lady will be when you come home tomorrow with a bunch of fresh Ben Franklins. You'll be a hero."

"No way, man."

I could feel the client sweating and trembling. Jack was babbling about craps and food courts and who knows what else.

We never did go to AC that night, but bull markets can be a lot of fun.

• • •

I walked into Steve Smith's office and sat down, waiting for him to get off the phone so we could go for a beer after work. He was babbling on the phone to someone and I gave him my best impatient look. He kept talking, then handed me the phone. I covered the mouthpiece and asked, "Who the hell is this?"

"Just talk, Mr. In-a-hurry," he shot back.

"Hello," I said into the phone.

"Oh, hi. This is Paul Carroll from the *Wall Street Journal*. I was just talking to Steve and he told me you might know something about semiconductors."

Paul Carroll was doing a story on IBM and Digital Equipment and needed a quote on the state of the memory chip business and where prices were going. I was happy to talk to him—Steve did me a big favor. The next morning, I was quoted in Paul Carroll's story, saying something half intelligent about memory. By lunchtime, reporters from both *Forbes* and *Fortune* had called. So had reporters from the *Dallas Morning News*, the *San Jose Mercury News*, the *Chicago Tribune*, *Computer Resellers News* and *Electronics Times*. I guess a lot of people read the *Wall Street Journal*.

• • •

Jack's stories of his boxing experience didn't hurt his reputation. Wall Street is infatuated with athletes. I think it is a macho, competitive thing. Or maybe it is something interesting to talk about beyond bond yields. There was Steve Ally, the hockey star. The salesman from Minneapolis had played for the Vikings. The bond-trading firm, Cantor Fitzgerald, hired most of the Cornell lacrosse team. Everyone you met was either a junior champion tennis player or a senior champion squash player. There were ex-Giants and ex-Jets scattered through the broker ranks. At a happy hour at the South Street Seaport downtown, I ran into Matt Doherty who worked somewhere on the Street. He had played with Michael Jordan at North Carolina, and would eventually become their coach. One of the sales traders at Paine Webber was once an assistant basketball coach at Syracuse.

And everybody golfed. Except me—I just killed worms. The salesman who covered Connecticut decided to have a golf outing at Stanwich, the ultra-exclusive country club in Greenwich. I was on an upswing as an analyst and was invited. The special guest was Rick Pitino, then head coach of the New York Knicks basketball team, who had been at Syracuse.

I was told to go change in the locker room. I could use any old locker, as they were all unlocked. The one I picked was Ivan Lendl's. It didn't look like it was used very often, because the crease in the plaid pants was too crisp. I left an Intel report in the locker, in case Ivan wanted to talk stocks. The greens all had doglegs.

With all those athletes, you can imagine the overdose of testosterone on the stress-filled trading floor. Traders were almost all street-smart. They had to be. But often they would go after each other and fights were common. There is an old Wall Street adage (there are so many good ones) that says the best way to find a great trader is to take a cab ride into Queens and hire the first three guys you see when the meter reads $10.

One sales trader from Queens named Dennis was the alpha male on the trading floor. Dennis had a big bulging neck and a massive chest that looked out of place poured into a tailored shirt with a tight collar, a yellow power tie, and bright suspenders with football helmets on it. If he didn't like what you were doing, like ripping his client off on a trade, he would walk over to your turret, prop his foot and massive leg up on your trading station, and start chanting J-E-T-S JETS JETS JETS JETS. Not everyone backed down, and blood would occasionally spill. All in a day's work.

. . .

Jack finally found *the fight* to go to. Mike Tyson was to fight Tyrell Biggs in Atlantic City, in one of his first defenses of his heavyweight title. Biggs had beaten Tyson during the Olympic trials and represented the U.S. at the Los Angeles Olympics. Apparently, Tyson never got over it, so this was a grudge match. The fight was in Atlantic City on a Friday night. Jack got tickets on the floor and a black stretch limo to take us there. Mike Lockwood, the block trader, was paying for everything. A sales-trader friend of Mike's came along, as did a friend of Jack's who worked at a small telecom company. I was number five.

We got stuck in traffic getting out of New York. Mike Lockwood and the sales trader were going on and on about how ugly the market was, and what a tough day it had been. I remember Mike saying, "It will be interesting to see what happens on Monday morning."

Mike didn't feel good about it. This trader talk is often dull. I didn't have a worry in the world. It was a bull market, my stocks were working, and I was headed to a prizefight on Friday, October 16, 1987. What could possibly go wrong?

CHAPTER 5

Time to Get Serious

The prizefight was great. We sat about ten rows from the ring. Mike Tyson won in eight or nine rounds. I immediately bought five red baseball hats that said Tyson-Biggs on them and the sight of five Wall Street types stumbling through casinos must have been strange. Then again, in Atlantic City, maybe not.

The five of us sat down at a blackjack table. The dealer took a look at us and asked, "You guys from Wall Street?"

"Yup."

"I don't know why you guys come down here," the dealer said as he shook his head, "you guys run the biggest casino in the world, up there in New York."

I suppose he had a point. Working on Wall Street during a bull market sure was fun. The night lasted forever, quite literally. I got dumped back at my apartment in Manhattan at 9 a.m. Saturday morning.

• • •

Monday was a strange day, to say the least. The market started going down immediately, and kept dropping. By lunchtime, it was clear that October 19, 1987, was going to be historic. It was just down, down, drown. By 2 p.m., most of the research department was lining the perimeter of the trading floor, just gawking. No one was having fun. The phones kept ringing. After a while, traders just stopped answering them. Over-the-counter traders would occasionally look up and yell "What the fuck are you staring at? This ain't a circus. Go away."

No one did. At the close, the market had dropped 508 points or 22.6%, on top of the 100 or so point drop the previous Friday that had spooked Mike Lockwood.

It was the end of an era, or so everyone thought. This pissed me off, because I was just starting to have fun. Trading over the next two weeks was like a yo-yo. Up, down, up, down, sideways, up, down. Ed Kerschner was constantly on the hoot and holler, the open mike, during the day.

"According to the KP model, the market is undervalued by 12%."

Twenty minutes later it was, "Now the market is overvalued by 7%. No wait, it's going down, it's undervalued by 3%."

Tom McDermott finally came over and told Kerschner to shut up.

Rumors swirled that banks were going bust. I stopped at the ATM on the way home and took out the limit.

The crash was more a wakeup call than the start of a depression, as many had feared. It wasn't a repeat of the 1930s. Wall Street did restructure. Layoffs started, but I can't recall any analysts being let go. Instead, more were hired.

Companies kept reporting earnings and the stock market

opened for trading five days a week. Salesmen needed something to call investors about. Institutions needed advice from analysts, or so I was told. Research seemed immune. Underneath the surface, however, changes in how Wall Street was paying for analysts would change the game in subtle ways. Some of it had to do with those over-the-counter traders not answering their phones. Wall Street got sued for not answering phones and the SEC insisted the Street put in a system known as Small Order Execution System, SOES. This automated execution of small orders would lead to day traders and would eventually lead to automated trading systems known as ECNs. These ECN trading systems would represent over half of over-the-counter trades by 2001, with commissions a hundredth of what they had been in 1987. It changed the way Wall Street gets paid. Commissions were toast. Banking fees would replace commissions, and eventually kill research in the process.

. . .

For the rest of the year, everyone was in shock. Jim Carroll, the #1 oil services analyst, came into my office, and he told Jack Grubman and me that he had just gone into Margo Alexander's office and told her that she could cut his pay, but to please not fire him. After Jim left, Jack and I decided that was not the greatest negotiating tactic.

Bob Cornell came by and told stories of bear markets in the past. His advice was to not freeze, and to keep working, as hard as ever. Bob went on, "When I worked with Ben Rosen, and Intel would blow up as it inevitably did, he would just freeze. The pink phone message slips would just pile up in front of him on his desk and he just couldn't pick up the phone and admit being wrong. He was the ax in Intel's stock, but couldn't take being wrong. So he froze. It probably shortened his career as an analyst."

Jack turned to me and said, "Yeah, and Ben Rosen turned into a centi-millionaire venture capitalist." Still, the "don't freeze" story was good advice.

One afternoon, I got into a conversation with Dennis Callahan, the insurance analyst, and a bunch of other analysts in the hallway. Dennis was telling us that, at a wire house, you should never recommend a stock that goes down.

Simple enough, but how did you know? This scared the hell out of me, since my stocks went up and down like the Cyclone roller coaster at Coney Island, often for no rhyme or reason. But he was right. Retail investors only listen to you say Buy or Sell, they don't bother trying to understand your underlying reason. Most institutional investors listen to your reason and didn't care about your Buy or Sell rating.

Trading was down. Paine Webber's retail brokers and their expensive offices (there were three retail offices in Milwaukee) were a drag on the business. The research department started cutting back on traveling, which of course, is the meat and potatoes of analyst marketing. All of a sudden, it was a liability working at Paine Webber. As I hadn't made *I.I.* yet, no one else wanted me, and I had to hunker down.

Jack Grubman would constantly mention, not just to me but within earshot of anyone and everyone in the department that he had just talked to Al Jackson, the director of research at Prudential Bache. There was no way he was going to move from one retail house to another, but this talk helped at bonus time. Steve Smith started complaining that Morgan Stanley had just hired one of his competitors, Carol Muratore, and hadn't even called him. Of course, he just wanted to go back to Margo and have her match the offer.

. . .

Aware they had a revenue problem, management at Paine Webber began to hire investment bankers to leverage the research department. Bob Pangia, who had previously run a technology group at Drexel Burnham, was hired out of Kidder Peabody. With his sidekick, Dave Lahar, and a former Baltimore Colts cheerleader, Susan Harmon, out in San Francisco, we began to scour Silicon Valley for deals. It was frustrating as Morgan Stanley and their technology banker Frank Quattrone were winning most of the good deals.

On one trip, Pangia and Lahar picked me up in the morning at the Santa Clara Marriott, and we drove around to four or five meetings. We ended up north of Santa Clara, near Burlingame, with no more meetings. Bob Pangia decided we should squeeze in nine holes of golf. We found a course and played in the mud—my Mo Ginsburg suit and shoes were caked—until around 2 p.m. We all had seats on the 3:15 flight back to New York. There was only one problem—my car was in the parking lot at the Santa Clara Marriott. "Tough shit," Dave Lahar told me, "that's out of the way, and we'll never make the flight."

"But what about the rental car?" I asked.

"Don't worry about it, Paine Webber owns National Car Rental. You can deal with it," he laughed.

When we got to SFO, I called National, lying through my

teeth that the car wouldn't start that morning, there were these exploding sounds, and they should go pick it up in the parking lot of the Santa Clara Marriott. No problem, the woman at National told me. Lahar and Pangia laughed the whole flight home.

When I got back to my office, I told my secretary to check with National to make sure they got the car. The next week, she came into my office with a funny look on her face and told me that National had called and there was a warrant out for my arrest on charges of Grand Theft Auto for the missing car. I hadn't planned on a trip to the West Coast that week, but got on the next flight to San Francisco, and drove my Hertz rental car down to the Santa Clara Marriott. I found the car, popped the hood and started pulling wires as quickly as I could. Then I called National, told them that the car was safe and sound, that it still wouldn't start, and they should send a tow truck to the fifth car in the second to the last row at the Marriott. The charges were dropped, but I haven't listened much to investment bankers since then.

• • •

Gunning for *I.I.* votes, I got myself into a frenzy by competing with the other analysts on the Street. My main target was Tom Kurlak, since he was the new #1 analyst and the most visible, but there were others in my line of fire. My competitive juices were running wild and I demonized each of them: Alan Rieper at Cowen was known as the Grim Reaper. John Lazlo at Morgan Stanley was Last to Know Lazlo. I despised them all.

I was trying to emulate Steve Smith, who was still the #1 computer analyst on Wall Street. He had one of those rare things, a Sell call, on Hewlett-Packard. The company just hated him. They wouldn't invite him to analyst meetings, forgot to alert him to conference calls, and would bad mouth him to institutional investors. To make matters worse for HP, Steve was right on the stock for a long time.

I had worked a summer at HP when I was in college, and still had some stationery and business envelopes embossed with Hewlett-Packard's logo that I had, er, borrowed. Jack Grubman and I spent an entire afternoon writing a scathing letter to Steve Smith from John Murtaugh, a senior executive vice president at Hewlett-Packard. We made up the name, along with everything else in the letter. Steve was a "cringing milksop" for saying Sell on an American institution such as HP. They would soon kill Digital Equipment, whom Steve was recommending, to become the most important computer company in the world. HP was considering filing criminal charges for "misanalysis" and was starting a campaign to have Steve "delisted" as a practicing analyst on Wall Street. It was pretty rude.

On my next trip to the Valley, I think it was the rental car wire-yanking trip, I mailed the letter from Palo Alto. Fortunately, Jack and I were sitting around when the letter arrived and was delivered to Steve. He came out of his office with the blood drained from his face, and stumbled around for a few minutes and then went back inside and shut the door. Jack went into Steve's office to see if he could "console" him. I went into my office, shut my own door, and dialed Steve, claiming to be John Murtaugh from HP. For some reason, Steve took the call. I started berating him, telling him I was going to "come on out to New York and personally kick your ass if you don't change your opinion on HP." This went on for a while, and I finally hung up when I saw Jack literally crawling on the floor back to his office laughing so hard that he was in pain. We never told Steve and I'm not sure to this day whether he had any clue it was Jack and I who were the fictitious John Murtaugh.

. . .

It was time for the next offsite research department meeting at Arrowwood. Margo announced that the night before the meetings, there would be a costume party. Come as your industry or come as your fantasy career. I went as a Japanese samurai. It worked for both criteria. Jack went as a boxer, complete with boxing gloves, boxing shoes, bright red shorts and a yellow cape. About halfway through the cocktail party, I borrowed padded gloves from someone dressed as a lacrosse player, and started boxing with Jack. It started playfully enough, but then I landed a left jab on his face and he decided to teach me a lesson. I covered up like Ali playing rope-a-dope with Foreman, but Jack just beat the crap out of me, with a sick grin on his face. I slid out of there as fast as I could.

We held our poker game/room destruction event, as usual. I ordered the customary 12 Heinekens and 45 Kamikazes and as they arrived, so did Margo Alexander. The name of her administrative assistant was Steve, and Steve and Margo came dressed as S&M. Margo's costume was complete, including chains, a leather dress, and a whip. It was a shock she was joining us, but Jack had invited her. Margo partook with the best of us. She was clearly having fun. Jack finally turned to her and said, "Margo, break something."

She smiled, took her beer glass and smashed it on the table. Jack smiled even more. Management was now complicit with our scheme. We could no longer get in trouble for wrecking the room; we were with management.

• • •

The tech group marketing effort kept going. We convened in Cleveland, Ohio, with plans for a huge breakfast the next morning with every big account in town. We all flew in from different places and met in the lobby of the Stouffers Hotel. The salesman who covered Cleveland was buying drinks. Jack Grubman was there, as were Steve Smith, Curt Monash and his assistant, Bob Therrien. I got there about 11 p.m. from San Francisco, and quickly caught up.

There was a commotion at the lobby bar where we were sitting. Two big groups were at the Stouffers that night. First, there was a mass of hairdressers in town for a convention, so there were lots of strange looking men and women with bad hair and wild outfits.

The other group also had bad hair and wild outfits, but was much more interesting. Boxing promoter Don King, with his hair pointing straight up, was in the next set of seats over from us. He had an entourage, of course, but sitting with him was none other than Mike Tyson, with his own posse. At midnight, I told Jack I would pay him $1000 if he went over to Mike Tyson and popped him in the nose. Steve quickly matched my offer, as did Jeff and then everyone else. Jack had $5000 coming to him to punch Tyson, whom we had seen in Atlantic City not one year earlier. Jack got a funny grin on his face, and said no way, he might be stupid, but he wasn't suicidal.

No matter how many times I called Jack a pussy, he refused to take a swing at the heavyweight champion. He did, however, go over and shake hands with Don King and Mike Tyson. I thought for sure he was going to offer to split the five grand with Tyson in exchange for a punch in the face. Jack was on his way to becoming a Wall Street heavyweight.

• • •

I had been sleeping in one Monday morning with nothing particularly important to do that day. I was getting a little sick of being an analyst, too much stress, not enough rewards. The phone rang right next to my head, jarring me awake. It was Steve Smith. "Where the hell are you?"

"Uh, I was sleeping until a few seconds ago, thanks for waking me up," I said.

"Get your ass in here, you just made *I.I.*, runner-up in Electronics."

It was the fastest shower I had ever taken. A $5 cab ride later, I was sitting in my office basking in the glow of the infinite wisdom of *I.I.* magazine. Steve was still #1 in computers. Jack had jumped to #2 in telecom, which meant his picture and a description of his calls were in the magazine as well. Runners-up only had their names appear. Bob Cornell called offering me congratulations. I gave him full credit.

Jack started going on about Al Jackson of Pru calling. "Yeah, yeah, I've heard. So has everybody else," I said to Jack to try to shut him up.

"Have they really?" he asked.

My phone rang. It was Adam Cuhney from Kidder Peabody, my competitor with whom I barely spoke. I still hated my competitors. "Hey, congratulations on the *I.I.* ranking. I'm moving over to investment banking but they won't let me move until I hire my replacement. Since you just made *I.I.*, how would you like to work here? We'll pay you whatever you want."

"I'm interested. Let me get back to you," I said.

This was my market offer, even though I knew it wasn't a very credible choice for me. Kidder had three things working against it. I knew they had no bankers because I was working with the ex-Kidder bankers. Second, they were in the middle of an insider trading scandal on their arbitrage desk, I remember reading about a couple of guys from their risk arbitrage desk who bet on takeovers and ended up being dragged out in handcuffs by the feds. And finally, they were downtown, on Hanover Square. That was way too far for me to commute.

However, Adam did offer me three times what I was making, so I went down and interviewed at Kidder one morning. I insisted on meeting the heads of research, sales and investment banking, to see what their philosophy was, and to see if I could stomach the commute for some place more interesting than Paine Webber.

I met with Mike Madden, who ran all of equities. I asked about the scandals—do they mean anything, and would he be around to see all this through. "No problem. I am here for the long haul. The arb stuff is isolated. This firm is stronger than ever."

Both Jack and Steve told me to stay away from Kidder, or any banking firm. I'd lose my credibility as an independent analyst, I would just become a whore to banking and every institutional client would know it and stop voting for me in *I.I.* They had proof.

There was a bunch of boutique investment banking firms, mostly in the Bay Area. Robertson Coleman Stephens, Montgomery Securities, Hambrecht and Quist, LF Rothschild Unterberg Towbin. In Baltimore, there was Alex Brown. Jack pointed out that none of their analysts made *I.I.*, because no one trusted them. They were banking analysts, always recommending stocks of companies their firms took public. Fidelity hated them. JP Morgan hated them. They are "skank" as Jack put it.

Margo found out about my Kidder interview and matched their offer. That was that—market offers work. Two weeks later, the *New York Times* ran an article about the recently fired Mike Madden and what a mess Kidder was. Of course, Kidder was eventually sold to General Electric, who put a manufacturing guy in to run it, prompting the famous line, "That's just what we need, a good tool and die guy around here." GE had their own problems with faulty internal reporting of trading profits by Kidder trader Joseph Jett, and they eventually sold the firm to Paine Webber. Shit travels in circles on Wall Street.

• • •

Curt Monash started showing up at our group marketing meetings wearing an arctic parka and sneakers and increased his condescending tone to clients. Jack decided that Monash was a liability to the tech group. It was true. He was an embarrassment, a moody brat with the personality of a 14 year old. Also, his *I.I.* ranking had slipped. So Jack had him fired. I'm not sure if this is what Jack really wanted, or if it was just to prove to himself that he could have someone fired.

He didn't think things out very far into the future, because Margo promoted Curt's assistant, Bob Therrien, to become the software analyst. Therrien had been a retail broker at First Jersey Securities, one of those sleazy bucket shops, and had related to me how First Jersey would recommend stocks to a Long Island branch and the next week, recommend a New Jersey branch buy them, and the Long Island branch to now sell them.

Therrien was not a particularly good analyst. His hot report after his promotion in 1988 was called "Broken Windows," on the demise of Microsoft. He also fell under the spell of Ed Esber, the CEO of the software company Ashton-Tate. One day, Bob ran into my office with stock trading charts of Ashton-Tate. "You see that? You know what that is?"

"No," I replied.

"That's a flag pattern. I can't believe it. This is so great. Flags are awesome."

"What does it mean?" I asked.

"When you get a flag, it means that the stock is either immediately going up or immediately going down. You never know which way, but it is going to move."

"Great," I sighed.

Therrien ran to Tom McDermott demanding time at the next day's morning meeting, Ashton-Tate's stock was going to pop, he told Tom. He got on the next morning, pounded the table on TATE, with a few lame reasons. He forgot to mention the flag pattern. The sales force went out with the call, and sure enough, by 10 a.m., the stock had popped from something like $10 to $13. Bob strutted around all morning. I really did have to figure out these damn charts.

We had a tech group meeting with the folks at JP Morgan Investment Management at lunchtime. The meeting went well, until Fred Kittler asked if anyone knew why Ashton-Tate was up three points that morning. Therrien stood up, put his arm straight out, formed a fist with his thumb pointing up, and then slowly brought his thumb to his chest saying, "That was me. I moved Ashton-Tate this morning."

"OK, thanks. Any particular reason?" asked Fred.

Therrien went on and on spewing nonsense about their products and the quarter and great management.

Two weeks later, the stock collapsed from $13 to $7. The flag was right, the stock was going to move, but it was going to move down, not up. Therrien had been the ax in the stock for two weeks, and then the ax chopped him up. Therrien later left the company, becoming one of those upside-down purplish figures flying towards Caesars Palace in the painting. I swore off charts forever.

. . .

I had a newfound respect, now that I was an *I.I.* analyst, but it didn't last long. You have to prove your worth every day, but with the market bouncing around, it was hard to get a grasp on a good long-term call. In the meantime, we won a piece of IPO business. I had met with Jim Diller, CEO of a company named Sierra Semiconductor. We had a great meeting where I tried to convince him that he didn't want to be a standard cell, custom chip company, that those were a dime a dozen. Instead, since half of his sales were chips for modems used for dialup access to on-line computers, he should position the company as a telecom chip company. He hemmed and hawed, but his company was growing like a weed and chose Paine Webber to lead his IPO. This was my first experience with a real live deal.

The timing was awful. I was getting married in Chicago, and flew to the wedding from New York by way of San Jose, California, to sit in on drafting sessions for the deal.

My wife and I honeymooned in Australia, about as far as you can get from Silicon Valley and Wall Street. We returned through San Francisco, and it took about five minutes to realize that in three short weeks, the technology industry was in the tank.

The stocks had cratered, orders disappeared, and companies were slashing their earnings estimates. I, of course, was stuck recommending stocks in the group, and was dead wrong. This was not the first time I was wrong, and it would not be my last. I just hoped it would be contained to 49%, and not spill over and ruin my be-right-51%-of-the-time thing. I dreaded the trip back to New York to face the firing squad at the morning meeting.

• • •

Strangely, one of the calls waiting for me on my return was from Rod Berens, the director of research at Morgan Stanley. They needed a semiconductor analyst. Their chip analyst, Last-to-Know Lazlo, had moved to banking, although perhaps not by his choice, is how Rod explained it. It didn't matter to me. It was like that sliding square game, with 15 tiles and one square empty. You just move tiles around to get them in order. Lazlo opened a square and I was asked to move into it.

This time I was interested. I was dead wrong on my stocks and facing a long ugly period rebuilding my credibility with the sales force. Paine Webber was going through its own changes. Retail was beefed up with the hiring of a bulldog named Joe Grano, who had done wonders at Merrill Lynch. We were all told to do anything Grano asked of us. Bob Cornell had moved to investment banking, but eventually left the firm and put his analyst hat back on at Lehman Brothers. Margo Alexander left her job as director of research, with a promotion to run equities. She was in charge of sales and trading. There went all my advocates. I dreaded the idea of retraining the new management team about my worth. Not an easy task when your stocks are down and out.

I met with Rod and his sidekick Peter Dale, who had both recently moved from trading at Morgan Stanley to fix research. I would be their first hire, and I was recruited hard. I told them I needed to meet bankers as well as the head of sales and equity before I could decide. I didn't want to be a whore to banking and jeopardize my *I.I.* slot. I flew to San Francisco and met with Frank Quattrone, who ran West Coast technology banking for Morgan Stanley. His office was filled with tombstones—framed ads from the *Wall Street Journal* that chronicled all the deals he had done. It was an impressive list of companies, the best of the Valley, although I do remember seeing one for Miniscribe.

Frank wanted to know how soon I could start, as he had some deals for me to work on right away. "I'm not sure, Frank." I wasn't sure of Morgan Stanley, let alone my starting date.

When I got back to New York, I met with Frank's boss, Carter McClelland. Carter asked, "How soon could you start? I've got a few deals that need you right away."

I got spooked. I told Rod I didn't want to be a whore to banking. "Don't worry," he told me, "that doesn't happen at Morgan Stanley. Instead, we have something unique to the Street, a meritocracy. If you do well, you'll get paid, whether it is with bankers, with great calls, or with *I.I.* rankings."

I liked the sound of it all, but was unsure about the offer. I wanted to stay pure as an analyst. What an idiot I was. Research was already changing. Soon there would be no more pure analysts, no virgins giving unconflicted advice to money managers. I needed someone like my old friend to tell me again, "What are you some kind of idiot? Just take the job."

Margo found out about my talks with Morgan Stanley, even though I didn't want to play the old market offer game again. I told her I had met with top management at Morgan Stanley, but was still unsure.

"You need to meet our top management," she told me. The next day, I was up in Paine Webber's CEO Don Marron's conference room. In it were original Ansel Adam photos, framed but not yet hung up on the wall. I found the one that matched the poster in my bathroom. That didn't help. Someone was making enough to buy this stuff, or maybe the firm was paying for it. Either way, I felt more like a lowly cog than an important part of the firm.

Don Marron came in and talked for about ten minutes. I forget what he said except for one thing. He was not ashamed that Paine Webber was terrible at investment banking. He told me that being an analyst was an important job and that I should stay independent of bankers, lest I become just another piece of "Wall Street Meat."

He was dead right. But my choice was staying his piece of

meat, so he could buy more hallucinatory art, or figure out how to stay unconflicted somewhere else, yet leverage myself via banking. If I could pull it off, I would be my own piece of meat.

I called Rod Berens the next morning and told him that as long as Morgan Stanley didn't have automatic light sensors in their offices, I would accept his offer.

CHAPTER 6

Wheeling and Dealing at Morgan

Morgan Stanley was two blocks south of Paine Webber on 6th Avenue. The commute was identical. I turned right instead of left off the subway.

Morgan Stanley's offices were anti-plush. No cherry wood credenza and matching chairs. The carpets had been installed when Kennedy was president. It seemed like Morgan Stanley had raided a Salvation Army store for their furniture. Rod Berens told me the company's philosophy was to generate returns and pay high salaries, so people could buy nice furniture at home. On my first day, I got my new office with a view down 6th Avenue that was almost identical to the one I'd had in my last job. It was as though I had never moved, and it was a little spooky. I was seated next to Paul Mlotok, the #1 oil analyst on Wall Street, who had also started that day.

About five minutes after I sat down, who should walk into my office but the Piranha himself, Hank Hermann. He was in New York and visiting with analysts to get some ideas. He started giving me shit right away, looking at my shoes, and saying he was checking for my white shoes, a reference to the snotty "white shoes" reputation of Morgan Stanley. I showed him my veins and told him my blood had mysteriously turned blue when I walked into the building. The snotty days were over, I figured, if Morgan Stanley was hiring me. Perhaps it *was* a meritocracy.

The next person I saw was Ed Greenberg, Jack Grubman's competitor whom I had met at that boxing dinner a year or so before. Ed was a little concerned. "You know, there is no way Morgan Stanley would hire Jack Grubman," he told me. "That kind of shit doesn't go over around here." I'm not sure what he really knew, but it was clear that Jack already had a funky reputation on the Street.

• • •

On my first day, there was a monthly equity department meeting. It was in a monstrous conference room, the same one where the morning meeting was held. It was on the 20th floor, the same as the traders. Anson Beard, who was most definitely a blue blood in addition to having married into the IBM fortune, got up and gave a "let's rile up the troops speech."

"Wall Street is changing, and we need to figure out how to change with it. It's hard to get paid for trading anymore. Commissions are now six cents a share and will be a nickel soon. We used to get an eighth or a quarter. Not anymore. We just hired two analysts today, Andy Kessler and Paul Mlotok. Stand up, you two." I snapped to my feet. "Kessler follows semiconductors, and Mlotok oil. Let's do some math. We are paying these two studs a bundle. At six cents a share, we'd have to do 70% of all the trades in oil stocks and 110% of all trades in semiconductor stocks just to pay these two guys' salaries, with *nothing* left for us. Go figure that one out. This firm had better come up with another way to pay these boys, or we're going out of business. Now get out there and figure it out. Let's go."

OK, it was not quite Patton addressing the Third Army, but it sure set the tone for my days at Morgan Stanley. I wasn't coddled or eased into my job. Instead, I had every sales person and trader knowing exactly how much I was getting paid and feeling like I was taking money out of their pockets. Great start.

I sat down with Rod Berens the next day. "OK, you got me into this mess. I've worked on Wall Street for almost five years and I have no idea how it works. Can you walk me through how firms make money and how everyone gets paid?"

"Not really, it is a mystery. But I can get you close."

So he started talking. Wall Street is about allocating capital. Great companies can get money easily—bad ones have to pay more for it. Wall Street gets paid by controlling access to that capital, and charging fees to get it. Companies pay via banking fees and investors pay via trading commissions. That's the easy part. The dirty little secret is that the people on Wall Street keep half of all the revenue they generate. If you look at Morgan Stanley's or anyone else on the Street's profit statements, you'll see that compensation is half of their expenses. Communications is around 10%. There is interest expense. There is overhead for buildings and lights. And whatever is left over is reported to shareholders of the firm as profit. But the biggest expense is compensation, the famous bonus pool.

If you are inside a Wall Street firm, the trick is to associate yourself with as much revenue as you can, and make your case for as close as possible to 50% of that as you can for yourself. That used to be easier to do. Most of it came from trading commissions. Before 1975, commissions were 75 cents a share. After the Big Bang that ended fixed commissions, they dropped to a quarter or an eighth (12.5 cents) per share. By 1989, when there were more firms on Wall Street with big trading operations, commissions were five or six cents a share. By 2003, they were often a penny a share.

Wall Street responded in a few different ways. Banking became much more important. You could charge a 7% fee for taking a company public. You could charge 2% to raise money for a company that was already public. Trading desks also started creating derivatives. Not pure stocks, derivatives might be as simple as a convertible bond or something more complex that is part stock, part bond, part warrants, and part option.

No one knew the real value of derivatives, so huge fees were hidden when the derivatives were sold and companies could make big profits trading these convertible bonds or derivatives. Traders would buy them when they were undervalued and sell them when they were overvalued, because only the issuing firm really knew what they were worth.

. . .

It turned out that Morgan Stanley sucked at research. The discipline that I had had at Paine Webber, of getting the Buys and Sells right, marketing to clients year round and making 100 calls a month, was nonexistent at Morgan Stanley. The morning meeting was short. There were so few good analyst calls that bankers would come in and pitch their deals. Reports took weeks to get out. Morgan Stanley did OK in *I.I.*, because there were still some remnants from Barton Biggs' All Star hiring from a decade before. But it was fairly lame. Barton left research to build Morgan Stanley Asset Management into a powerhouse, but without him, research deteriorated.

This turned out to be great for me. The sales force was hungry for interesting calls. They would take me to see any client I wanted any time I wanted, and no one else was going. Rod Berens and Peter Dale had promised to clear out the deadwood—Mlotok and I were just the start of their hiring.

• • •

The market opens for trading five days a week. In holiday weeks, it is guaranteed to be open for four days. Even during Thanksgiving, there are never two days off in one week. Companies report earnings once a quarter. But stocks trade about 250 days a year. Something has to make them move up or down the other 246 days. Analysts fill that role. They recommend stocks, change recommendations, change earnings estimates, pound the table—whatever it takes for a sales force to go out with a story so someone will trade with the firm and generate commissions. For some reason, Morgan Stanley was into price targets. I hated them. To me, they were pure marketing fluff.

I would recommend Intel at, say, $25. The first question I would get is what is my price target. My answer would be $40 for no particularly good reason. It was high enough to interest investors, but I was guaranteed to be wrong. If it hit $38, it was a great call, but I was wrong. If it went to $60, it was an even better call, but I was still wrong.

What usually happened was that if the stock hit $35, I was

asked to adjust my price target to $50, so that the sales force would have a call to go out with. This was how you got target creep, an ever-upward bias on numbers so calls could be made. It would come back to bite Wall Street by the end of the decade.

• • •

I stuck around New York for the first week, then was on the shuttle to San Francisco to drive around with Frank Quattrone. He immediately started calling me Dr. K., which was the nickname of Dwight Gooden, the strikeout king for the New York Mets. I wanted to call him Q-Man or just Q from James Bond fame, but everyone was doing that. So instead, I decided to call him Frankie. He was a local boy done good out of Philadelphia, like Jack Grubman. He was also the worst dresser on Wall Street. He had a bushy mustache, a $3 Casio watch, ties from Kmart and there was always some rip or tear in his suits. He also was smarter than shit. Within five minutes, he could tell whether a company was going to be a franchise player five years down the road. It was uncanny how right he was.

The first company we went to see, along with Steve Strandberg, another banker who worked for Frank, was Cirrus Logic. This was a dime-a-dozen custom chip company with no real value added, run by a guy named Mike Hackworth. I told Frank this on the way out of the building and he said he thought the same thing—that it wasn't Morgan Stanley material, and he wanted to see if I agreed. Hey, a team player. I was no whore to banking.

That also impressed me. Morgan Stanley had a quality filter. They wouldn't take just anybody public, do a piece of business just for the sake of generating fees, but they would rather find the best of the best. I liked that.

"What else do you have going on, Frank?" I innocently asked on the way out of Cirrus's parking lot.

"I can't tell you."

"Top secret or something?" I asked.

"Hasn't anyone told you about the Chinese Wall?"

"The what?"

Frank went on to explain that there was a separation of research and investment banking in Wall Street firms. He didn't like it, but that's the way it was. Maybe a banker was working on a takeover deal. He couldn't go telling the analyst about it. There would be too much of a temptation to turn around and recommend the stock. Material and non-public information stayed with the bankers. Analysts had to be left out in the cold. The Chinese Wall, was a conceptual separation, so that bankers and analysts could work at the same firm and ease concerns about the use of insider information. It sounded like an excuse for Frankie to keep me in the dark.

Our next stop was to a company called Xilinx, to see their CEO, Bernie Vonderschmitt. Xilinx was an interesting little company that made programmable gate arrays, like the first company I followed, LSI Logic. Engineers could program the chips themselves rather than pay hundreds of thousands to LSI Logic. I liked the company, and I liked Bernie. I started telling Frank this but was interrupted. "We are already set to file the deal when the cycle turns. I just wanted Bernie to meet you and agree that you were a good enough analyst to follow his company."

That popped my balloon. A week on the job and, as predicted, I was already a piece of Wall Street Meat. I just couldn't let that happen again.

• • •

I didn't want to waste any time on the new job. My industry was in the toilet, orders were terrible, and the stocks were beat up. I wanted to put a Buy rating on my stocks as fast as I could, and start marketing to clients. The conduit to the morning meeting was Richard Dickey, the salesman. He was a barking dog, just like Tom McDermott, and put on an angry front to make sure that analyst calls were worthy. Unfortunately, that scared most analysts away. I was used to a barking dog and wasn't intimidated. Richard and I became fast friends. He had been at Morgan Stanley for several years, and knew the history on everyone.

I quickly got to know the rest of the sales force. They were my conduit to clients. The guy who covered Boston, which had tons of *I.I.* voting accounts, was Doug Quartner. On a shuttle ride up to Boston, Doug asked me "What year?"

"Huh?" I mumbled.

"What year were you?"

"What year was I what?" I asked.

"What year did you graduate from Harvard Business School?"

"No year, I didn't go there," I answered with an annoying tone. "I don't have an MBA." I noticed Doug spitting up his coffee. "Actually, I'm an electrical engineer, with a bachelors from Cornell and a masters from Illinois."

A strange look came over Doug's face and finally, with teeth clenched, he mumbled, "Morgan Stanley doesn't hire people with backgrounds like that."

• • •

In April, four or five weeks after starting at Morgan Stanley, my boss Peter Dale invited me to Opening Day at Yankee Stadium. It was earnings season, and there were a few conference calls I had to be on because, being at a new firm with a demanding sales force, I didn't want to fall behind what was going on in the industry. Naturally, I accepted. You only live once.

Also going were Morgan Stanley's biotech analyst, Michael Sorell, and Jack Mueller, a new salesman just hired from Smith Barney. A stretch limo picked us up on 50th Street, and wound its way to the Bronx, with only a brief stop at a deli on 3rd Avenue for a case of Heinekens, "in case we get stuck in traffic." This was going to be a long afternoon.

It was a beautiful day, and Yankee Stadium was packed. When I looked around our seats, which were in the fourth row behind the Yankees dugout, it seemed like Wall Street had emptied out and filed into the Bronx. There were lots of paisley ties and suspenders. Having grown up in New Jersey, I had been to Yankee Stadium a million times as a kid, sitting in the bleachers or in the nosebleed upper decks. I had never sat in the Bob Eucher "frunt reuwww" section.

Peter Dale made sure the beers kept flowing. About the sixth inning, I whipped out my Motorola Star-Tac cell phone and dialed the number for the Advanced Micro Devices conference call. I put the phone on mute and like a dork, listened to the call, even asking a question during the Q&A section. I was diligent.

The Yankees were awful, and similar to the stock market, they were at the beginning of four dismal years. On the limo ride back to Manhattan, we stopped at a bodega somewhere in Harlem for more beers. This was not the brightest thing for four lily-white guys in business suits to do. I was volunteered to run in and fetch the beverages. I bought two 12-packs of Molson Golden Ale. A jar at the counter caught my eye, so I made the impulse purchase of pickled pig's knuckles as well.

Back in the limo, I insisted that in exchange for me getting the beers, everyone had to take a bite of the pig's knuckles. I got no takers, no matter how loudly I prodded. Clouded by a day of beers, baseball and better than expected earnings, and to prove I was more macho than the other three weenies in the limo, I took a bite. Big mistake. Fortunately, the moon roof was open and most the day's joy was spewed out along Adam Clayton Powell Jr., Blvd.

• • •

I had recently moved to Westport, Connecticut. My wife and I had decided we couldn't raise children in Manhattan, and we liked what we saw in Westport. The only problem was a commute of an hour and twenty minutes versus a quick subway ride. Since I was in the Bay Area so often, I didn't have to commute every day, but what was hard were the dinners in New York. While the sales guys at Morgan Stanley didn't have the contest of running up the bills, there were still lots of dinners, either with clients, or with traders and salesmen. I would always check my watch around nine or so, worrying about which train I was going to take home. They were every hour until ten or so, then one at 11:30 p.m. and nothing until 1:30

a.m. At one dinner I sat next to Jack Mueller, who caught me looking at my watch. "Where do you live?" he asked.

"In Westport. The trains are kinda funny," I said.

"I'm in New Canaan. Don't worry about it, you and I are on the late train, the 11:30 p.m. I call it the Vomit Comet."

The train was filled with drunken commuters stumbling on to make the last decent train home. Sure enough, someone that night helped remind me of the train's nickname, all over his shoes.

• • •

In May, I was off to Europe. Morgan Stanley had a huge office in London and others in Tokyo and Hong Kong. International travel was part of the job, although not quite a perk. In Europe, I was off to visit 14 cities in eight days. One of them was Dusseldorf, Germany. A friend of mine asked if I was going to bomb the ball bearing factories. "The what?" I asked.

"The ball bearing factories. Every episode of *Hogan's Heroes* is about bombing the ball bearing factories in Dusseldorf."

Of course, Dusseldorf was a thoroughly modern city, having been leveled by Allied bombing—to get the ball bearing factories, I suppose. The London-based salesman, David Trenchard, took me to the old city of Dusseldorf, which was filled with beautiful cobblestone streets and dozens of beer halls. We met up with one of his clients from a big German bank. After many beers, David turned to the client and said, "Tell Andy the Christmas party story."

"No, you tell him."

So David Trenchard launched. He had been invited to the

bank's Christmas party up in their offices. Steins were flowing and the party quickly got out of hand. Since they were sitting by the trading screens and phones, David's client picked up the phone, called Morgan Stanley's trading desk and put in an order for 10,000 shares of a stock. I think it was Eli Lilly, but it may have been someone else.

The trader asked if he had a price limit. "No, buy it at the market." Filling the order made the stock go up half a point. This amused the client, who was trying to impress David. He called to back New York. "Buy another 50,000 shares, no limit."

The stock jumped 1¼.

"Buy another 50,000."

"OK, now buy another 50,000." By this time, the entire party was watching and laughing hysterically. The stock was up 3½.

Soon, there was an asterisk next to Eli Lilly's symbol on the screen. They pulled up the news story. Dow Jones Newswire reported heavy buying of Eli Lilly by a "foreign investor, perhaps a German pharmaceutical company." Now the party was screaming with delight.

A few minutes later, another story appeared. The New York Stock Exchange had halted trading in Eli Lilly, on rumors of a takeover by a large foreign company. This put a damper on the Christmas party—they couldn't trade any longer. Everyone went back to the punch bowl and the client quietly sold the position the next morning. I learned a funny lesson. Stocks do talk when they move up or down, but often they are liars.

• • •

Morgan Stanley had a glorious history of technology analysts, led by analyst Ben Rosen. But that history was long ago. The

Morgan Stanley I joined had a rag-tag technology group. The first person I spent time with was the computer analyst, Carol Muratore. She was the one who got the job instead of Steve Smith. I was hoping to work with her as I had worked with Steve Smith. But it didn't take long to figure out that she was mailing it in, and hadn't lifted a finger to do research for quite some time.

The next person I thought I could work with was the software analyst, Jim Mendleson. Richard Dickey filled me in on his background. Jim had been Ben Rosen's assistant when Ben was the #1 electronics analyst on the Street. Ben Rosen had brought in the Apple IPO in 1978, and was an early proponent of the importance of software. Back then, there really weren't any software stocks. Rosen had left in 1981, apparently in a huff for not getting paid for banking deals, and Morgan Stanley didn't know what to do with Jim Mendelson. Then spreadsheet maker Lotus went public, and whoever ran research thought, "Hey, Jim, you were Ben Rosen's assistant, so you must know something about software." It was sort of a battlefield promotion. Probably half the analysts at Morgan Stanley were promoted assistants.

Jim followed mainframe software companies, like Cullinet, that don't exist today. By 1985, personal computer sales were growing, Lotus had gone public, and Microsoft was considering picking investment bankers to lead a potential IPO. This was a layup for Morgan Stanley, as Carter McClelland and Frank Quattrone had built a strong technology banking franchise.

Founder Bill Gates and President Jon Shirley of Microsoft were making the rounds in New York, meeting with investment bankers, in what is known as a beauty contest or a bake-off. Morgan Stanley put together a formidable team, including Carter, Frank, CEO Dick Fisher, and the heads of sales, trading and syndicate who managed the deals. Oh yeah, and Jim Mendleson was there as the analyst who would cover Microsoft. The meeting went well. The pitch book showed how Morgan Stanley was the leader in doing technology deals, with page after page of tombstones and stats to prove it.

When the meeting hit a lull, Carter McClelland turned to Jim Mendleson and asked, "So, Jim, what do you think about PC software?"

Jim immediately jumped in. "As you know . . ." Uh-oh, whenever an analyst starts out with as-you-know, it means they are clueless and digging up some thought from the past to prove their credibility.

"As you know, I am not a big believer in personal computer software. I am much more sanguine about the prospects for the mainframe software segment, with PC software most likely to remain a niche for the foreseeable future."

This was his marketing pitch to institutional investors who owned Cullinet and other names that would end up being trashed by Microsoft. Carter couldn't cut Mendelson off fast enough. Gates and Shirley left the meeting, went downtown and selected Goldman Sachs to lead their IPO. Mendelson was honest, but he was dead wrong and literally put Goldman Sachs in the technology banking business.

Miraculously, Mendelson didn't get fired. Years later when I arrived, he was still around and following Microsoft with a hold rating. As you know, he was still not a big believer in personal computer software.

Then there was Rick Ruvkun, a squirrelly little guy who was

afraid of his own shadow. He was hired to cover any of Frank Quattrone's deals that fell through the cracks or didn't fit in any particular category. Rick was afraid to recommend any stocks, afraid that Richard Dickey would yell at him, afraid that clients would make fun of him if they went down. Steve Strandberg, the investment banker who worked with Frank Quattrone, had even written one of Ruvkun's reports.

Rounding out the group was George Kelly, an ex-IBM salesman who had heard you can make a lot of money on Wall Street and insisted he knew a lot about networking companies. There were only a few, mainly those that made modems from the chips that Sierra Semiconductor produced. George quickly figured out where his bread was buttered, and practically lived in Frank Quattrone's office out in San Francisco. His big break came when he worked on the Cisco IPO. He later insisted that he personally brought the deal into Morgan Stanley, but that's not the story that Frank Quattrone told me. George really was meat for bankers. So be it, it made his career.

I was on my own. There was no Steve Smith or Jack Grubman to learn from anymore.

. . .

On a marketing trip to Boston in 1989, on the two coldest days in February, I learned my most valuable lesson yet. Boston was a tough town. The people there took a backseat to the Wall Street types who worked in New York, but between Fidelity, Putnum, State Street Research and Mass Life, there was an enormous amount of capital being managed in Beantown. And in some Red Sox-Yankees grudge thing, they loved to beat up on analysts.

Doug Quartner had packed the schedule tight, and there were eight to ten meetings a day where the same things were said. At the end of the second day, I found myself on the 40th floor of yet another anonymous Boston building, meeting with my client at the Boston Company. I was exhausted and in a bad mood, but the client was in a worse mood.

"Why is it that you like Intel?" he asked.

"Well, the stock is cheap, and if you look at . . ." I started to reply. I was interrupted immediately.

"Wait a second. Cheap? What do you know about cheap? Listen son, you have no right to come in here and say something is cheap or expensive. You have no idea how the market works. You, you're nothing. You're just the pulse of the market. I gotta figure out what's in a stock. You're just a tool I use. You and all those other stupid analysts are just the consensus. The market values stocks every day based on your dopey inputs. If I can figure out how you are wrong, and the consensus is always wrong, I can make money. But I need to take your pulse. If I think your estimates are too low, because of some new product or some upturn in the economy, I can buy the stock and make money. So don't come in here and tell me cheap. Just tell me what you think about the company's fundamentals and how much you think they are going to make, this year and next year. I'll figure out if the stock is cheap or not. Thanks for coming. Meeting ended."

• • •

I think back and laugh at how stupid I must have seemed pitching a growth story like Intel and semiconductors back in 1989. Leveraged buyouts, LBOs, were all the rage. Junk bonds, despite the bad name they got leading up to the '87 crash, were plentiful. Salomon, Goldman, Merrill and even Morgan Stanley had taken share from Drexel Burnham peddling high-yield bonds to whoever could use them.

Technology companies couldn't use these junk bonds. Research and development to advance technology cost too much, and either you pay interest or you fund R&D. Anyone back on R&D would be dead within 18 months.

Of course, I was still new to Morgan Stanley as an analyst. For months, I was pounding the table recommending Intel's stock, but the stock just went sideways. Almost everyone at Morgan Stanley was convinced that I was a worthless buffoon. Then, in October of 1989, the strangest thing happened. United Airlines was trying to go private, in a huge, employee-led leverage buyout, but the deal started to unravel. Just like that. Junk bond buyers refused to play, somehow assessing the risk to be too great. The junk bond market collapsed, all in one day. The Dow dropped 190 points, and everyone at Morgan Stanley, myself included, fell into a deep mental depression. This was a mini-crash, but it reminded everyone of the real crash two years earlier.

That day, for some strange reason, Rod Berens bought a case of champagne and threw a party in the hallway of the research department to celebrate. Now, I'm not one to turn down champagne—Margo Alexander at Paine Webber had bought it every 100 points on the way up. But now we were down almost 200 points. The party was just outside my office, so I had no choice but to attend. When I walk out, Rod marched over to me, gave me a high five and whispered in my ear, "It's your turn."

I had no idea what he was talking about, but he was dead right. The market gave up on buyout names and spent the next 12 years focused on growth, led by Intel, Microsoft, and companies like Cisco and America Online that were not yet public. The 1990s growth-over-value era clearly started on that champagne fogged day in October 1989, and not a moment too soon. Hey, another lesson. Market transitions are always violent, but spell opportunity if you can figure out the change that is taking place.

• • •

I didn't know about any opportunity except that I was recommending a bunch of stocks that were dead until an up cycle hit the industry. Bob Metzler, the head of sales, would often walk past me after I was on the morning call and mumble, "Cycle, my ass."

Of course, that just made me work harder to prove one was coming. I also spent lots of time out on the road. On a par with the "Sherman's March to the Sea" marketing trip, most analysts' least favorite trip was known as the "Ohio Death March," an action-packed 36-hour adventure. Jack Mueller was the salesman in the territory and led the march. First, we flew to Columbus, Ohio, for an 11 a.m. meeting and then a lunch meeting. The trip continued with a quick flight to Cincinnati for afternoon meetings.

It turned out that the Bengals and the Browns were set to play Monday Night Football at Riverfront Stadium, right near our hotel. Unfortunately, Jack Mueller had set us up for a dinner in Louisville, about an hour's drive from Cincinnati. We agreed that if we made it back in time, we would go to the game. "You don't need tickets to get into these things, you know," I told him.

The Louisville dinner lasted until nine, and we ran out the door and into our car to race back to Cincy. But first, we badly needed beers. Seeing a huge roadside sign that said BEER, we pulled into a drive-through package goods store. Upstairs, we noticed a dozen pool tables, and Jack Mueller, who was driving, insisted we stay in this inviting store for a while. "Drive-through beer sales, pool tables. C'mon, it doesn't get much better."

"Jack, the battle of Ohio, Monday night football. Please, stay focused," I insisted.

We rushed back to Cincinnati and ran for the stadium. It was 10:30 p.m. and almost halftime.

We walked around the outside of the stadium trying to figure out how to get in. Unlike events in New York, there were no scalpers. At one point, Jack Mueller had his head sticking through some steel bars, trying to slip inside. No dice.

We finally came to a ticket window that was open and we ran up. I said, "Hi, we're from Morgan Stanley and are here for our tickets."

"Who?"

"Andy Kessler and Jack Mueller."

"From where?"

"Morgan Stanley."

"I mean, where is that?"

"New York City."

"Did you say *New York Post*?"

"Yes, that's it," Jack added.

"And your name?"

"Jack Mueller."

"Did you say Jim Miller?"

"Yes, that's it."

"OK, here are the tickets. Enjoy the game."

Enjoy we did. Cleveland won, and we were back at our hotel by 12:30 a.m. We had a 6:30 a.m. flight to Cleveland, so we had to be up by 5 a.m. and out the door by 5:30 a.m. But Jack Mueller still had a hankering for pool, so we slogged around Cincinnati until we found a bar with a pool table as well as Heinekens and Kamikazes. My head hit the pillow at 3:00 a.m.

It was a rough plane ride to Cleveland, ouch. We made our 7:30 a.m. breakfast, I started my pitch about the semiconductor cycle, and Jack Mueller babbled about the Browns and pool. The client was eating very runny eggs. That didn't help me at all, as my brain felt as runny as the eggs. We hurried out, drove three hours to Pittsburgh, had a lunch with several clients, and flew home in time for me to make a Motorola earnings announcement back at the office. Death march, indeed.

• • •

You did these trips to make *I.I.* I couldn't help but notice that my competitors at the technology boutiques were not *I.I.* ranked. But they were doing lots of deals. Robertson, Alex Brown, Montgomery, and H&Q had decent analysts, but no one on the Street trusted them. They were banking analysts, and always recommended the shares of companies they took public. Since that was their role, their lack of objectivity was understood. These firms were partnerships that gave up objectivity to get paid. Institutional investors understood this, and didn't use them for research, other than explaining the latest deal. They were obviously conflicted, but so what? The interesting small deals that these analysts brought might someday turn into something big, maybe even the next Microsoft. But clients certainly didn't vote for these analysts in the *I.I.* polls.

At Morgan Stanley, Frank Quattrone had a dilemma. He wanted to be king of the boutiques and have his pick of the best of the Silicon Valley deals. Goldman Sachs could eat his crumbs, and the boutiques would do all the rest, the really skanky deals. The Morgan Stanley quality filter would reign. But, the situation became increasingly difficult as these boutiques grew and added analysts in narrower and narrower disciplines like computer-aided-design, peripherals and database software.

So-called "bulge bracket" firms like Morgan Stanley (named because their names were in the top row and printed with the largest type fonts on the cover of a prospectus for a deal) were organized into *I.I.* slots. As analysts did not want to follow companies where there were no votes to be had, you stuck with the big names. Frank Quattrone complained to me that when Peter Dale called disk drive companies "periph-ree-als," it was clear he was never going to get the analyst he needed.

Frank also ran the Mendelson risk: that an analyst might blurt out that we are not believers in some new market or another. Frankie engineered around the problem by explaining in his pitch books for deals that Morgan Stanley Technology Banking was a "boutique within a bulge bracket firm." Analysts would be intimately involved in the process. The entire middle section of the bake-off presentation would be by the analyst, who positioned the company in a way that had been agreed to ahead of time. This had two purposes. First, there were no Mendelson-like surprises. Second, it promised the company a ranked analyst who worked closely with the bankers to help ease the company into public life. Somewhere later in the presentation was a timeline that showed the analyst initiating research coverage 25 days after the IPO. It didn't promise a Buy rating or favorable research. It didn't have to. Of course, the analyst liked the company. The analyst was sitting there smiling away, hoping to bring in the deal so he could make the case for a bigger bonus at the end of the year. As Anson Beard pointed out at that first equity department meeting, analysts weren't going to get paid for trades they generate. Banking was it.

This was a brilliant move on Frank's part. It absolutely killed Goldman Sachs in the market. Goldman couldn't win a mandate to do an IPO for a decent company to save their life. They did Microsoft, because of Mendelson. They did Sun, I think because venture capitalist John Doerr at Kleiner Perkins an investor in Sun, wanted someone to balance Morgan Stanley and Frank Quattrone in the Valley. But Goldman didn't do much else of quality, until the late '90s. Even then, they had to relax their standards to win IPO mandates.

It may have been a brilliant move, but it was the beginning of the end for research. Even though I worked for Morgan Stanley, I was increasingly pitched as a boutique analyst. But boutique analysts were not trusted by institutions. This was a very subtle shift.

• • •

January was bonus time at Morgan Stanley. Most professionals on Wall Street draw a salary, and are paid an annual bonus that is often several times the size of the drawn salary. In January, management would do reviews, figured out how the firm did, how big the pool was, how much you contributed to the firm, and then how much it took to keep you at the firm and not jump to a competitor. If you were one of those that didn't contribute much, it was time to pretend you did. Each January, I would see Jim Mendelson dressed up in a fine suit, tailored shirts, with bright new ties, and a gold pocket-watch dangling from his belt loop. He would get up and walk the halls any time he saw the director of research or anyone from management walking by. "Good morning, sir." "Good afternoon, sir."

Jim would often plop down in my office and lament how hard his life was. "It's tough to make it these days. I've got my home in Westchester and a maid and a gardener and a nanny. That stuff runs up. I've also got my place out in the Hamptons and a maid and a gardener there as well."

He didn't get a lot of sympathy from me. I figured the guy was going to be fired any minute, but he was a Wall Street survivor.

Carol Muratore, on the other hand, continued to be missing in action. Most of the sales force confided in me that I ought to have her fired. I agreed, but I didn't run things. They pushed hard and I had chats with Peter Dale and Rod Berens about how it would be nice to have a computer analyst and how, despite having Carol Muratore on the payroll, Morgan Stanley didn't have one. They agreed, but their hands were tied. No one gets fired from Morgan Stanley. "Why not?" I'd always ask.

"Too messy, too much liability," they would say. It was a meritocracy, but one with tenure. Crazy.

Muratore eventually left the firm "to pursue other interests."

. . .

As did every firm on Wall Street, Morgan Stanley had a strategist, Byron Wein. Here was another guy who had outlived his usefulness. Byron was a nice old man, whose shtick was that he had been around the block a few times and knew everything. Like all strategists, he had a proprietary model that would say if the market was over-valued or under-valued, but Byron never really told anyone how his model worked. I think he rolled dice in his office. To be fair, I am biased against most strategists' top down work. You just can't model the stock market, because there are too many inputs changing too quickly.

Byron had this nasty reputation of taking another analyst's work as his own. "I am adding Intel to my recommended list today." Great, it's already been on mine. If it worked, he took credit, if it didn't—it was my fault, the perfect hedge.

Byron was also famous for his year-end, top ten predictions for the coming year. He would bill the list as bold and provocative predictions that had only a 30% chance of happening. He claimed his success rate was 70–80% each year. It took me a while, and then I finally figured out how he did it. They weren't so much predictions, but instead a series of conflicting statements. Something like, "The stock market will go up 10–20% next year, unless S&P earnings disappoint and then the market will be flat to down." This statement is *always* true, even though it contains a bold prediction of an up 20% market. The guy was a walking hedge. I accidentally had a hedge in my first report on LSI Logic, with a Hold on the cover and a Buy inside. This guy did it on all his calls.

Over the years, I have read Byron's quotes in the *New York*

Times and the *Wall Street Journal,* and he hasn't changed a bit. He continues the never wrong double-speak, and is almost always right, because he never actually takes a risk by saying anything. Maybe I was just jealous, as I actually had to say Buy or Sell and was judged with every tick of the tape.

. . .

I did OK at bonus time, but was told point-blank that I needed to do more banking work. Fortunately, things were about to turn up. In February, I had visited a chip distributor out on Long Island, and was informed that they had just gotten the numbers that afternoon. February's orders were tracking 10% ahead of January. It was lucky timing, but it was a sign the cycle was turning.

I begged Richard Dickey to let me on the morning call. He was skeptical, because I had said it too many times before. I told him I had real evidence this time, not just concepts that things were going to get better. I got on the call and exclaimed, "The cycle is coming. The cycle is coming."

The sales force went out big with the call. There wasn't much else going on. My stocks were all up 10–15% that day. All tech stocks were up 5–10%. The guy who writes the "Heard on the Street" column for the *Wall Street Journal* called and said he heard that I was "the ax" in Intel and Motorola and that my comments moved the group. While I took the credit, of course, and was quoted in his column the next morning, in reality it was the news of better orders that moved the stocks. But, who was I to set his perception straight? By finding the sign first, I was an ax, now I had to work on confirming it.

There is nothing like getting quoted in the *Wall Street Jour-*

nal to improve one's standing *inside* the firm. Bob Metzler grabbed me in the elevator the next morning and said, "Great call." I had finally arrived at Morgan Stanley! That weekend I was quoted in Barron's, and my counterpart in the Tokyo office told me I was quoted on the front page of the *Nikkei Keizai Shimbun,* their financial paper of record.

Timing is everything. February was too late for Morgan Stanley's bonus review. Jim Mendelson was probably paid more than I was with his dress-for-success formula. I didn't care. I was having too much fun. The up cycle meant that there was banking business to be had. Rod Berens was right. It was my turn.

CHAPTER 7

Up, then Tanked

Our first deal was Xilinx, that programmable chip company. Of course, it was Frank Quattrone's deal, not mine, but I jumped in feet first. It was March 1990, and Steve Strandberg and I wrote the prospectus and figured out how to market the deal. The burning question was where to price the stock. There is always tension inside an investment banking firm when it comes to pricing IPOs. The banker, who is charging a 7% fee to do the IPO, wants the deal priced as high as possible. The sales force wants to see their clients make money on the deal, and wants it priced as low as possible, so it trades up.

Since I wasn't going to see a penny extra from the deal, I rooted for a low price, in order for my clients to make money. Over the phone in my family room on a Saturday afternoon, Steve Strandberg and I agreed that the Xilinx stock should be priced from $6.50 to $8. Steve thought that was too low, that the board would never agree, but I figured it would mean strong demand for the IPO, which would send the price up.

After a prospectus or red herring (so named because of a disclosure written in red ink) is distributed to investors, the company embarks on a two or three week road show to pitch the story directly to institutions. I went on the road show with Xilinx management. Because it was my clients they were pitching to, I could help by making introductions and sticking around for a while after the meeting to tell investors what a great company it was. Of course, it worked well for me too, as it was an easy entry into clients' offices. I didn't have to say much, but if the stock worked, I would get credit by association. Plus, it didn't cost a thing. All expenses were charged to the deal. That's why you saw lots of limos, private planes, and expensive dinners during road shows. Shareholders ended up paying. My favorite five words were "Charge it to the deal."

Before a deal is done, every Wall Street firm has a commitment committee that must agree to it. The bankers have to present the deal to a group represented by the top guys from sales, syndicate, research, and banking. It is a system of checks and balances to ensure there is a check on quality. At firms like Morgan Stanley, it is by no means a rubber stamp.

Frank Quattrone called me up and told me to "lobby" the Xilinx deal to some members of the commitment committee, to make it go over easily at the actual meeting. I wasn't sure why he didn't make the calls himself. Later, I found out about his reputation of overpricing deals to maximize banking fees at the expense of investors. Because of this, salesmen hated Frankie.

I walked into Bob Metzler's office, and asked him if I could go through the Xilinx story. He gave me an annoyed look and asked, "Is it alive?"

"What do you mean?" I asked.

"Look, I can't be bothered with these piss shit little deals. I've got bigger things to worry about. If they have a pulse, if you say they're OK, I'm in. Now get out of my office."

The deal went over well. It got its commitment. After the road show, the final arbiter of the price is a syndicate guy, who doles out the shares, "feels" what the market might bear, and decides on the price. The Wall Street adage about syndicate managers is that they are too lazy to sell and too dumb to trade.

The price of $6.50 to $8 was a steal for this company. The book was 30 times oversubscribed, a record that would not be surpassed until Netscape five years later. This meant investors wanted 90 million shares versus the 3 million the company was selling. What had really happened was that investors smelled a "hot deal," meaning an IPO that would trade up after it was distributed. They over-ordered it and when the syndicate guy sensed this demand, he set the price at $10.

That morning I hung out with the over-the-counter trader, Tony Kiniry, who would trade Xilinx. Traders usually wait until an hour or so until after the market opens to begin trading an IPO, in the lull of mid-morning. There was never a lull. All morning, sales traders ran up to Tony and dumped buy tickets on this desk, which he sorted, flipped through occasionally and then figured out where to open the stock. Xilinx opened at $17 and never traded near $10 again. Tony turned to me and said smiling, "The ducks are quacking."

• • •

A few months after the Xilinx deal, Ed Greenberg came by my office. He had with him someone he introduced as Dan Reingold. Dan was, how should I put this, vertically challenged. He stature was Napoleonic.

"I'm moving into banking," Ed told me. "I've hired Dan, who I'll train as an analyst and then cut him loose. Dan was the director of investor relations at MCI, so he knows a lot about the Street. But if you can, help him too."

I shook his hand. He was a nice enough guy.

Ed added, "Dan, Andy came here from Paine Webber, and he knows Jack Grubman."

Dan jumped in, "You worked with Jack Grubman? Wow. He's a real prick."

"Nice to meet you too, Dan."

A couple of days later, I ran into Jack at Ben Benson's.

"Hey, how's it going?" Jack asked. "You a banking whore yet?"

"No. But I did meet a fan of yours, Dan Reingold. Greenberg hired him to take his place as telecom analyst."

"Yeah, I heard they hired that no-nothing clown from MCI."

"Reingold told me that you were a real prick."

"Well, you go back and ask Reingold if he still shops for clothes in the children's department."

Ah, great to see Jack hadn't changed.

• • •

The way to be a hero at Morgan Stanley was doing deals, either bringing them in or getting them done. My old colleague from Paine Webber, Steve Ally, the hockey star with the Wisconsin accounts, had moved over to Morgan Stanley just before I did. Now he had all the big accounts in Chicago and Milwaukee.

Morgan Stanley loved him and every other salesman was in awe. As the junk bond market disappeared, all the big LBO firms had to find other sources of financing for their portfolio companies. They had to de-leverage their companies—it was the 1980s in reverse. KKR, Kravis Kohlberg and Roberts were the LBO kings. They hadn't done much with Morgan Stanley, but approached their bankers and implied, "Look, we have a bunch of huge deals we are going to do over the next five years. But we have some dogs we need to get funded quickly. If you can sell these pieces of shit, you can do the easy ones with big fat fees later on."

The first dog was Owens Illinois, a glass and plastic company they had acquired in the heat of the LBO market in 1987. It was a howling dog, at that. The prospectus was ugly, the growth prospects dim, and the price too high on the cover. No matter, it was a live deal to be done and there was the promise of others. Steve Ally jumped into action. Most salesmen couldn't place a single share, but Steve placed over half of the deal with his accounts. He single-handedly got the deal done, and became an instant hero.

The next dog was Safeway, the West Coast supermarket. Investors laughed when salesmen called to sell shares in their IPO. There were no takers. Again, Steve Ally placed over half the shares himself. No one knew how, no one dared ask. Once Safeway was public, Morgan Stanley had 25 days before they were allowed to send out a research report on the company. The retail analyst wrote a lukewarm report, with a weak Buy rating on the stock. The bankers freaked, and threatened to fire everyone involved if the report wasn't more glowing. It got spruced up a little, but kept its wishy-washy "we're not so sure about this stock" feel. It was a battle between analysts and bankers that would play out again and again on the Street. The bankers usually won.

• • •

My cycle call was working out great. Through the summer of 1990, there was a rip-roaring bull market for technology stocks. Most analysts at other firms, especially the boutiques, were running conferences. Software conferences, computer conferences, networking conferences, semiconductor conferences. Clients were conferenced out.

I hated conferences. They seemed like a lot of work and with everyone doing them it was impossible to differentiate. But I got a lot of heat at Morgan Stanley. Here was a bulge bracket firm with no tech conferences. Jim Mendelson ran a software conference with, you guessed it, mostly mainframe software companies.

So I hit up on the idea of a reverse conference. Invite twenty of the top technology investors in the U.S. (there weren't many more), rent a bus in Silicon Valley, and drive around visiting the top companies. It was a lot less work, unique, and it earned you credit from the clients who really counted.

My rules called for a casual dress code. In addition, no management team was allowed to get more than three slides into a presentation before being rudely interrupted and pummeled with questions that got to the important issues. Of course, we never told the management teams this. They showed up in suits and ties, for the "Wall Street crowd" and were usually startled and delighted that smart investors were beating the crap out of them to get to exactly what was important.

It was a two-day trip and a keg was secretly smuggled onto the bus in the afternoon. A few of the companies we piled into were Intel, Sun, Apple, Silicon Graphics, and Adobe. The first day went well, and there was a buzz on the bus about what a great trip it was. That evening, there was no planned dinner. Instead we took everyone to Marriott's Great America amusement park across the street for chili dogs and roller coasters, a combination that didn't quite agree with Rick Ruvkun, last seen running towards the men's room. We then headed to the bumper cars, and took them over. I can't tell you how much fun it is to ram a client who rarely returned your phone calls into the boards with your bumper car, and have them laugh about it.

The next morning I woke up excited about the second day. I came down to the buffet breakfast we had laid out, to see everyone glumly poring through the papers. The headlines were stark. It was August 2, 1990. Iraqi tanks had rolled into Kuwait.

Just as I was having fun again, all bets were off. It was a new world beyond anyone's control—life on Wall Street, and everywhere else, just reset back at zero.

CHAPTER 8

Kicking Off the '90s

A few people on the bus trip packed up and headed home. The rest slogged it out, far less lively than on the first day.

When I got back to New York, Morgan Stanley was in panic mode. The world had changed. Oil futures had spiked. New trends were threatened. Inflation was going to return and the short-lived bull market of the year was in jeopardy. We were headed back to the '70s, or were we? Markets hate uncertainty more than any actual outcome.

But wait, we had the #1 oil analyst on the street, Paul Mlotok. I sat next to him. We started out on the same day. Surely, he would know exactly what was to happen to oil supplies and oil production.

Circumstances had made Mlotok the new superstar of the department. Meetings were arranged quickly with Mlotok, and attendance was mandatory. I filed into a conference room with almost every other analyst on the research floor. Mlotok handed out a table from his spreadsheets. On the left was current production by country. On the right was potential production based on their known exploration and drilling capacity. On the bottom was current worldwide demand. I don't remember the exact numbers, but demand was something like 50 million barrels per day. Current production was 50 million barrels, but 10 million disappeared when you removed Iraq and Kuwait. OK, what about potential production? If you added all the extra production from Saudi Arabia and the U.S. to current numbers, you got to no more than 45 million.

As if he were at Alliance Capital in Minneapolis, Mlotok cut to his conclusions. "Supply is less than demand, you can see by the numbers. The world doesn't have enough oil production to make up for Iraq and Kuwait, so oil prices are going higher, a lot higher. Your industries are all going to suffer, and mine is going to do well."

Steve Roach, Morgan Stanley's U.S. economist, known affectionately as the Roach Clip, had been forecasting a weak economy all year. Steve had a great track record, but loved to be a contrarian, forecasting doom when everyone was bullish. So he had been wrong until August 1990, but was soon heard around the department saying, "Thank goodness for Saddam Hussein. He got my forecast to work."

Mlotok laid out a pretty nasty outlook. This was to be company policy. Mlotok's oil price forecast and its effect on the economy was to be run through all analysts' outlooks for their industries and companies' earnings estimates.

If anyone had actually listened to Mlotok, the next morning meeting would have been filled with analysts cutting their ratings to sell. A few analysts did. I was spooked, but decided to keep my ratings. Good thing.

Mlotok was dead wrong. I mean absolutely, 100% wrong. He had had a chance to shine, but instead hogged the limelight, and went down in flames. The economy went into a recession, mainly because business travel dried up and companies cut capital spending in an uncertain time. Oil prices only briefly spiked at the end of September, when talks between Iraq and the U.S. broke down. Saudi Arabia had three times as much potential production as in Mlotok's table. They increased output and oil prices started heading down. It would take until January 1991 for all of this to play out.

• • •

In the interim, tension was high. A great year was ruined and the bonus pool was going to reflect it. People were short-tempered and nasty.

Someone had the brilliant idea of taping a monthly meeting of each industry group and what analysts thought of their segments and stocks. A transcript would be made, edited down to a few pages, and published. There were just four of us in the tech group, Kelly, Ruvkun, Mendelson and me. As far as I could tell, Ruvkun had no self-confidence, so he overcompensated by verbally beating up on people, without much reason or backing for the verbal tirade. He would often do this at these taped monthly meetings. At one in particular, he was just an argumentative stiff, disagreeing with everything I said.

The next day, I had a meeting with Barton Biggs, where I told him that what Morgan Stanley really needed was a technology strategist. He told me the firm hadn't had one since Ben Rosen, and that I should go ahead and just do it. I then voiced my frustration about the group and the fact that we needed to upgrade. Barton nodded.

Encouraged, I said, "This guy Ruvkun is a liability," lifting the words from Jack Grubman talking about Curt Monash years before. "He has baseless arguments and, at yesterday's meeting, I felt like reaching across the table, lifting up Ruvkun and slamming him against the wall to get him to shut up." I had, at least temporarily, turned into Jack Grubman.

Barton looked at me strangely, and ushered me out of his office saying, "Now *that* would have made for interesting reading in the transcript."

• • •

In January 1991, the dreaded earnings season had arrived, that two-week period from the tenth to the twenty-fourth that came around every three months. It was when companies reported their sales and earnings, and stocks flew around based on the numbers. It was an intense two weeks for an analyst, who had to instantly analyze the numbers, get a verbal comment out quickly to the sales force, listen to a conference call for more information, then write a two or three page report on what happened, how the numbers supported or disagreed with your stock recommendations. It was all in the spin.

For some reason, Intel and Motorola, the two names I was supposedly the ax in, reported on the same day, just after the market closed. They had coordinated their conference calls so that Intel had one for 90 minutes, and then Motorola immediately followed. I had a buy on both stocks. My earnings estimates were the high numbers on the Street for both companies.

Both companies reported weak results. After all, there was a recession going on, and they completely blew the numbers.

I wasn't about to pull my Buy ratings, so I spent most of the evening rummaging through their earnings releases looking for good news, new products—anything to put a happy face on a disaster. Both stocks were going to get hammered in the morning, but I had to go on the morning call and defend them. This is an analyst's worst nightmare. Maybe I should have listened to Paul Mlotok.

To make matters worse, there was a research department dinner that night, and I was so caught up in my own disaster that I completely forgot about it. I just blew it off. George Kelly, who liked to read corporate political tealeaves, told me later that I would have sat next to Barton Biggs. Instead, there was a nametag with my name and an empty seat all night. What a dumb ass I was.

Some time that evening, Operation Desert Storm began, available in living color on CNN. It didn't matter. I still dreaded the morning meeting.

The next morning, I got in early, headed to the trading floor prepared to get beat up, and the place was hopping. Bob Metzler had a big smile on his face. John Havens, who ran sales, looked at me and said, "It's a fucking high tech war. We shine lasers that guide bombs right down chimneys and blow the shit out of everything. It's already over, we won, and this market is going up, up, up."

Sure enough, stock market futures, which trade before the stock market opens, were up big, more than 3%. The market was going to pop. Intel and Motorola added five points each that day. The bull was back.

My mood went from glum to gleeful, with only a brief skipping of a heartbeat when Barton Biggs came up to me and said, "Hey, we missed you last night."

• • •

Sierra Semiconductor, the company I was going to take public in 1988 at Paine Webber until the industry downturn killed the deal, was back. The company went through three tough years, but was now positioned to grow. Cisco had gone public in February 1990, making George Kelly's career. Post-Iraq, tech was hot. Anything related to networking and telecom was heating up fast.

Following the Frank Quattrone plan, I helped position the company, and then pitched the deal at the bake-off. The whole "boutique within a bulge bracket firm" thing worked as it was supposed to, we got the business.

I joined the company on the IPO road show, as expected. Several times, I was the only person from Morgan Stanley traveling with the company. There were lots of deals in the spring of 1991. The co-manager on the deal was Montgomery Securities, a boutique notorious for its fierce traders.

During a rare break in the road show schedule in San Diego, we stopped for coffee. Sitting with me at an outside table in the beautiful sunshine was Sierra's CEO Jim Diller, their CFO, Steve Cordial, and the banker from Montgomery, Clark Gerhardt.

Jim Diller said, "OK, let's just do a hypothetical. I'm not saying this is going to happen, but let's say it's the end of the quarter, we add up our numbers, and we're short. We are going to miss the guidance we gave analysts. What do we do?"

I chimed in first. "What I would do is call your lawyers, quickly draft a press release that explains the shortfall and change in guidance and get it out as fast as you can." Some lawyers would disagree with my answer, saying you should just wait until the earnings release date and drop a bomb. I always hate that, since a salesman or some customer knows something and the stock starts mysteriously heading down.

Jim Diller nodded politely but he didn't miss many quarters, eventually merging with Pacific Microelectronics to create PMC-Sierra, one of the great stocks of the 1990s.

• • •

With the IPO market hot, Frank Quattrone was anxious to get more analysts to cover more industry segments. As Muratore was gone, there was no computer analyst. He was tired of using Ruvkun, and knew he couldn't win any deals with him. Mendelson was no help with PC software because, as you know, he wasn't a big believer. I was asked to see if I could cover PC hardware and PC software, in addition to Intel and friends. I said I would think about it. This was a suicide mission. I had a comfortable *I.I.* slot in electronics, and was moving up in the rankings. I was probably more valuable to the firm if I kept just that. Plus, Barton Biggs had agreed I should be a technology strategist, even though, or especially because, I hated strategists.

I gave it a try anyway. I joined Jim Mendelson at a Microsoft analyst meeting at their headquarters in Redmond, Washington. It was fun being at an analyst meeting without the hassle of actually having to pay close attention, care about earnings forecasts, and the like. I could relax, take it all in, and learn.

I recognized Rick Sherlund. He was the Goldman Sachs software analyst and ranked #1 in *Institutional Investor.* It helps when your firm takes the industry leader public, because you can almost become the de facto ax in the stock. This was the same advantage George Kelly would leverage with Cisco. It always pissed me off that I had to fight for that role with Intel, which went public when I was 14 years old.

Bill Gates spoke for a while about new versions of Windows. Version 3.0 had been introduced the year before and was a big hit. Other executives such as Frank Gaudette, spoke about applications and DOS and all sorts of interesting stuff. Then Jon Shirley got up to speak. "I want to comment about your earnings estimates. There are certain analysts out there, and you know who you are, whose numbers are just TOO HIGH. They have got to come down."

Slowly but surely, the room started emptying out. Sherlund ran out first, one-by-one, each of the sell-side analysts left the room, went outside, pulled out their cell phones and called their trading desks to tell them that Microsoft was talking down numbers. Jon Shirley finished his presentation to two people, Chip Morris, a buy-side analyst from T. Rowe Price, and me.

Analysts started filing back into the room, and word spread quickly that Microsoft's stock was down 4½ points. There were a lot of ticked-off analysts whose Buy ratings were wrong that day. Shirley finished and more Microsofties got up to talk about all sorts of minutia. I got up to go to the bathroom, and as I stepped out of the presentation room, there were Bill Gates and Jon Shirley standing there laughing as hard as they could. I heard Gates say, "What suckers, this is too much fun."

Microsoft was probably pricing their employee stock options the next week, and the timing of the analyst meeting was fortuitous to bang the stock. They played Wall Street like a fiddle. Of course, the stock has doubled so many times since then, and the 4½ point drop was just noise, but not on that day.

• • •

At the end of 1991, Rod Berens left Morgan Stanley. I think he had just about enough. He had transformed the department, or half of it anyway, from a bunch of sleepy analysts into a decent corps of respected analysts. There were still only five or six the sales force used, but that is the same at every firm.

I had lost my advocates. I still had Barton Biggs, who was a biggie at Morgan Stanley, as a fan, but I wasn't sure how far that would take me. Paul Brooke, the firm's drug analyst who had been there forever and knew the politics cold, came into my office and said four words, "Biggs has no coattails." Great conversation starter.

Without an advocate, people start picking on you. Morgan Stanley was sort of management by fire. The only carrot was the year-end bonus, but the other 364 days were filled with lots of sticks, i.e., people beating you up to do more for them. Research is a cost center. It generates zero revenues. So people who do generate revenue constantly bug you. Frank Quattrone bugged me, just about every day. That was OK, up to a point. I wanted to do more banking deals because I figured that was how I was going to get paid.

Jay Cushman, an elderly insurance analyst, took over as director of research for the U.S. Jack Curley, a manager who had bounced around the firm, ran research worldwide. Cushman's nickname was Moe-Larry, as in Moe-Larry and Curley of Three Stooges fame. They were both old school.

The other change at Morgan Stanley had to do with the downfall of Salomon Brothers in a Treasury bid-rigging scandal. A guy named John Mack had run bond trading at Morgan Stanley, somewhat successfully, but in the shadow of the trading power of Salomon Brothers. When Salomon went down, Mack went to Morgan Stanley CEO Dick Fisher and asked for capital, and lots of it. He said he could literally buy Salomon's predominant market share in bond trading.

Salomon was crippled, and Mack had to act fast before Goldman Sachs and Merrill Lynch stepped in. Sure enough, within only a few months, he was making a mint in the bond-trading department. He quickly expanded his fiefdom as wide as he could, placing his "people" in key roles around the firm. Just like that, trading was as important, perhaps even more important, than banking.

This was annoying. One day, it was decided that analysts needed to be reached at all times day or night by, traders and pagers were issued to analysts. Kinda like cowbells. As mine was handed to me, I ceremoniously placed it in my garbage can. Later that day, my secretary asked me why my garbage can was vibrating. I took it out and shoved it at the bottom of my desk. Jim Mendelson would later be seen sporting his pager when prancing around during bonus time.

• • •

Frank Quattrone ran a ski trip to Deer Valley in Utah every year. Every technology investment-banking client of Morgan Stanley was invited, as were those Frank wanted as clients. I would always hear the bankers talk about it afterwards: ski lessons with Olympian Stein Erickson (bend ze knees, $5 please), ski valets, great wine at après ski, oysters on the patio for lunch. It sounded like a great trip. So I asked Frankie if I could come.

"Sorry, can't do it," Frank said. "We want to stimulate open discussions with clients, and if analysts were there, it would ruin that openness. No one would want to talk and say what they really wanted to about their business and banking plans. Sorry, wish I could let you. Chinese Wall, you know."

Damn, I thought, what good is it working with these bankers if I'm not getting paid for deals and not even enjoying any perks? I caught Ed Greenberg in the hall, and was lamenting to him how unfair this was. He started giving me the "that's how it is at Morgan Stanley, the bankers rule," speech, when Alan Sebulsky, a healthcare analyst, walked by. Overhearing what we were talking about, he jumped in with, "Oh that boondoggle? I went on that last year. It's a great trip."

I didn't get mad. I just called Frank's secretary and booked myself on the trip, and told her that if Frank asked, "Tell him Alan Sebulsky told me I could go." I didn't miss another Deer Valley trip, without "bending ze knees" to bankers.

CHAPTER 9

Something about Mary

I worked with Frank Quattrone to find a decent analyst to cover Microsoft, Lotus, Compaq, and Apple. He wanted various analysts from boutiques. I helped interview a few, and they were all lame. Frankie accused me of not wanting to bring in anyone who was better than I was. Yeah, maybe. But I pushed for someone out of industry, who knew about technology and could figure out where the industry was headed. We didn't need another dull know-nothing analyst, an "analette," another Ruvkun. Of course, Frankie just wanted a piece of meat to cover his deals.

Then one day, Moe-Larry (sorry, I mean Jay Cushman) walked down the hall, followed by a woman I thought I had recognized. Without stopping to say hello or introducing us, he walked right by, pointed to the office next to mine, told the woman "This will be yours" and returned with her to his office.

Hmmmm, I wonder what's going on. The next Monday morning, this same woman poked her head into my office and said, "Hi, I'm Mary Meeker. I'm just starting today. How do you get a PC around here?"

My response: "Welcome aboard." For some reason, that was the official greeting for anyone new at Morgan Stanley. People say "Welcome aboard" non-stop, and you can't wait for someone else new to start so you can start saying it instead of having it pounded in your ears.

Mary Meeker came from Cowen, a so-so boutique that had a decent technology presence. She was an assistant to Michele Preston, the somewhat outdated computer analyst that I'd first met at the Apple analyst meeting on my first trip with Bob Cornell.

I was slightly ticked that the powers-that-be would hire a technology analyst without even asking me, but I suppose I had turned a few too many people down. I decided I wouldn't hold it against Mary. She seemed nice enough and bright enough, and eager to establish herself on the Street. She was also clueless.

No one asked me to, but I made the decision to work with her like Steve Smith, and Jack Grubman had worked with me. Give some advice, coach when appropriate, and take her on the road to meet clients and learn the ropes of analyst marketing. It would be nice to work with someone who knew what they were talking about again.

I offered to help with her first report. She was a little apprehensive of ANY help, but agreed her first piece, on Microsoft I think, could use some outside comments. It wasn't ready for primetime, or so I thought.

"Mary, this is very readable," I told her, "but the trick to an analyst report is to explain to investors something they don't already know. Figure out what everyone else thinks, and then explain why your outlook is different, for better or for worse. You've got to talk about consensus, and why your view is different. What do you see that nobody else sees? The stock is only going to go up if the company grows faster than everyone thinks. Are margins going up? Are there new products that will accelerate growth? Are there competitors that are going away? Lay out the case and then you have something to market to investors for years. All you've done so far is tell me how you feel about Microsoft's products. That's cool, but it's not enough."

It was true. Mary was a good writer, but it was all very emotional and not terribly forward looking. She was telling investors that she could "feel" that this stock was going to go higher because she had a good "feeling" about their prospects. That's fine for therapists, but not for analysts.

Mary and I started traveling around together, to tradeshows like Comdex and the Consumer Electronics shows, and to see clients. A group lunch in Cleveland, one-on-ones with Fidelity in Boston. Clients got to know Mary, and would tell her they would "feel" the same way as she did and they liked the feel of her reports. Hey, whatever works.

I would often tell people that I used to carry Mary Meeker's bags, so a lot of folks thought I worked for her. But I really did carry her bags. I would show up at JFK for a four-day trip with a briefcase that had my laptop, cell phone, three shirts, three pair of underwear, three socks, a razor and a toothbrush. Mary had three or four heavy bags. I'd announce immediately that there was no way I was going to carry any of them. It wasn't so much that there were lots of clothes. I had the suit on my back and she needed more of a change of clothes. I understand that. Her bags were stuffed with books, papers, computers, batteries and all sorts of crap that I'm sure she never used. Often, we would be running for a plane, and I'd look back and see Mary slowing down because she was as loaded down as a pack mule and fade into the crowd. I'd curse, run back to grab her bags of bricks and we would make the flight. It happened every time.

. . .

Mary got off to a terrible start with the sales force. It takes time to figure out how to pitch stocks to this group. The "touchy feely" stuff is no good. Salesmen can't call up clients and talk about their feelings. Imagine the bite the Piranha would take out of you with a call like that. Plus, the first few stock picks blew up in Mary's face. She quickly earned the nickname Heat Seeker. Her stock picks were like heat-seeking missiles finding the next stocks to blow up. To her credit, she ignored the criticism. She found sanctuary with Frank Quattrone.

Frankie now had just what he needed—an analyst with a growing reputation and no baggage (conceptual baggage, anyway). Frankie took her in to see any and every interesting new software company. Intuit, the personal financial software company, was a perfect fit. Private companies liked people who told them how they "feel" about their company. They needed an advocate when they went public, and Mary's shtick sounded good.

Richard Dickey would call me up and ask, "Where the hell is Mary?"

"I'm not her keeper. I think she's on the West Coast calling on banking clients," I answered.

"Well, tell the Heat Seeker that she'd better get her ass in here and defend those shit stain stocks of hers. They're a disaster and clients are bugging everybody for comments."

I told Mary that if she got the sales force right, it would be a lot easier with clients. But the warm comfort of investment banking had its lure. I could almost hear the calm, comforting words of Frank Quattrone telling Mary not to worry about those nasty guys in sales, because they were just monkeys with phones. The path to success at Morgan Stanley was doing deals. I didn't realize it at the time, but Frank was dead right. I was trying to point Mary in the direction of a traditional analyst. How old fashioned, how quaint. I was becoming old school. The Street had already changed.

• • •

Analysts didn't get paid directly for deals at Morgan Stanley. One year, they actually put in a formula to do just that, but someone in management wised up and put a halt to it. They didn't want a paper trail back to the formula, so banking compensation remained a loose concept. Bankers had an input into my year-end bonus. A good or bad word from bankers could add or subtract 25% of my pay in a heartbeat. Obviously, when bankers called, you listened. With John Mack's folks permeating throughout the firm, and trading becoming more important, bankers got religion. They couldn't let traders, the deep-in-the-heart-of-Queens types, take over the firm. Most traders at Morgan Stanley had MBAs, but no matter, the bankers got aggressive. They were driven to generate revenue, which had higher margins than trading. So bankers began calling on every little thing they thought they could get fees from.

Although Morgan Stanley had a good garbage filter, it seemed its standards were dropping ever so slightly.

There were a few technology bankers in New York. Half of them reported to Frank and the other half reported to folks in New York. I made the mistake of taking a few days off during the week in the summer and was driving around town doing errands. I got home and my wife, Nancy, was clearly agitated. "Some guy has called our house a bunch of times demanding to speak to you. He started yelling at me when I said you were out. He says you've got to be in Boston tomorrow or you're in big trouble, and could get fired. Who is this guy, and why is he yelling at me?"

Tom Eddy was one of those New York bankers. The boom came down on him to check out a company named American Superconductor near Boston that was considering going public. Superconductors were in the news, and were perhaps the next hot investment. I thought they were a scam so never really looked deeply into it. He called me and told me that Venrock, the venture capital arm of the Rockefeller family, was an outside investor in American Superconductor and needed the company to go public immediately so Venrock could stop throwing their own money into it. Venrock pulled some strings at the top of Morgan Stanley, and the heat trickled down from there, to Tom Eddy, to my wife, to me.

"Tom, I'm on vacation," I told him

"Not anymore, you're not. You'll be in Boston tomorrow at ten. This is from the highest places at Morgan."

"You're kidding right?" I asked.

"Nope."

"OK, I'll take the morning off from vacation and go to work. Did you have to yell at my wife?"

"Oh, sorry about that."

So it was off to Boston. As it was a beautiful summer day, I drove. I met up with Tom outside of American Superconductor and told him he owed me, big time. "Yeah, yeah," he said.

We sat through a bunch of presentations that sounded hokey. Then we got a tour of their factory. It turned out they make wires. They put some magic material inside some rubbery gunk, and stretched it until it became a mile long wire. This company was right out of the 1920s. The factory probably was that old.

"Tell me about the superconducting material," I asked.

"Well, it's still work in progress, and will probably take a few years to get working."

"How many years?"

"Maybe three, maybe five, maybe seven."

I leaned over and whispered to Tom, "Get me the hell out of here." We got up and left 15 minutes later. In the parking lot I told Tom, "There is no way we are doing this deal. This is a jam job by Venrock."

It was worse than the KKR Owens Illinois or Safeway. There was nothing there but gunk. Tom agreed and reported back to his boss that we had done thorough due diligence and the company might be ready for an IPO in three years.

A month later, I noticed an announcement that some fifth-tier Wall Street firm was taking American Superconductor public. My wife wasn't amused when I showed it to her.

• • •

A few weeks later, Frank Quattrone called. "Didn't you work on computer-aided engineering at Bell Labs?"

"Sort of, not really," I told him.

"We have a bakeoff tomorrow morning for one of these CAE companies, named Synopsys. They are picking bankers at the end of the day. Call Bill Brady with some positioning thoughts and then get out here on tonight's flight."

"This better not be another American Superconductor."

"Trust me, it's not."

I got a call on my cell phone at SFO from Brady asking me for more positioning thoughts. At the hotel the next morning, I got another call from Brady asking me for more ideas. "Have you been up all night?" I asked.

"Of course. I don't sleep much when the IPO window is open."

I met up with Frankie and a bleary-eyed Bill Brady in front of Synopsys. An even more bleary-eyed banker named Kristen Turner came running up carrying a dozen pitch books. Frank was calm. "Dr. K, check out the positioning section and figure out something coherent to say quickly. And tell them you are excited about their company, and will follow the stock. The usual drill."

It went well. "Boutique in a bulge bracket firm" and all that. We came, we pitched, we exited.

In the lobby, we passed a group from Goldman Sachs. I recognized John Levenson, the #1 computer analyst who had unseated my old pal, Steve Smith. Eff Martin, who ran Goldman's tech banking group, was there. Also present were three of the obligatory bleary-eyed bankers. Frank said to Eff as he walked by, "Don't even bother, you can save your breath. We already won the deal."

He was just kidding, but after I flew home that afternoon,

there was a voice mail from a tired-sounding Bill Brady saying we had indeed won the deal. No "thank you for coming out" or anything like that. Just a matter-of-fact, we won it. I didn't get paid anything for this deal either.

Frankie owned Silicon Valley and was the Goldman Sachs killer.

• • •

I didn't get paid for any deals at Morgan Stanley. Jack Curley told me "Don't feel so bad. Ben Rosen brought in Apple Computer in 1978, and he didn't get paid for that, or any other deal." So I had heard.

Of course, Ben Rosen left a few years later, in a huff, for not getting paid for banking deals. He did go on to become a successful venture capitalist, the chairman of Compaq computer, and added lots of zeros to his net worth. Note to self: see if this whole Ben Rosen thing is worth emulating.

• • •

Mary Meeker kept doing deals, which enhanced her reputation with institutions. They figured she knew more about Intuit, for example, than anyone, and would call her to find out how she felt about it. This was very different from the type of marketing I was used to. She stopped talking to the sales force almost completely. She would pitch deals, and occasionally change her earnings estimates, but she was not a call that the sales force used. Still, her reputation from banking made her name with clients. I told her this was a two-edged sword. As long as the deals were good, which they were, she would shine. But quality was out of her control. She told me she didn't feel that was the case.

"Mary, the trick to being a great analyst is to be skeptical as hell. Don't take anybody's word for it. Trust, then verify," I told her.

"My companies are usually right." That was classic banker talk.

To Frank Quattrone, I was Dr. K., but Mary was Dr. Feelgood.

• • •

I often wonder if that was the better way to go. Probably once a day, I would have a salesman call and yell at me for something. Richard Dickey explained it to me. When a stock goes down, the portfolio manager at a mutual fund who had bought it can't take the blame. So he calls the brokerage firm that "sold" him on the stock idea and rips the shit out of him on the phone. It's OK, that's part of the job.

Salesmen get paid for getting yelled at. After a while, however, a salesman can take only so much and eventually picks up the phone and reams out the analyst for the bad call. You have a thick skin as an analyst, work hard to never be wrong, or stop making any risky stock calls at all, so you won't get yelled at.

John Huller was one of the up and coming salesmen at Morgan Stanley. He was in charge of a group of sales people, in addition to covering big accounts like JP Morgan. He desperately wanted to be a hero like Steve Ally, to make a deal come together, get a big block trade, and do record commissions with a big account. But since none of this happened, he played the politics game, trying to get involved with decisions, always sticking his nose into things to try to get credit.

One afternoon, bored to tears, I went up to the trading floor to hang out with Richard Dickey and David Boucher, another top salesman. We talked for less than a minute and Richard said, "Watch this. John Huller is going to come over here and talk to you about your stocks and how sales can be more effective and is there anything he can do to help." Then Richard and David disappeared.

"Hey Andy, got a second?" It was John Huller. He had sneaked up behind me.

"Sure, what's up?" I asked.

"Tell me a little about your stocks. I want to see how sales can be more effective in communicating your call. Is there anything I can do to help you in your job?"

I looked over to the other side of the trading floor and there were Richard and David bent over laughing and pointing and looking away when John Huller followed my gaze. I talked to John for a few more minutes then walked over to the two laughing hyenas. "What gives?" I asked. "How did you know he was coming over and exactly what he was going to ask?"

David Boucher filled me in. Huller had gone to the head of sales, Bob Metzler, and demanded a promotion and more responsibilities. Metzler told him that he had no people skills and that until he could prove to Metzler that he had them, and could help analysts and the like, Huller would never be promoted.

It's a funny story until you realize that no one who has been at Morgan Stanley a while ever really gets fired. They just shuffle people around to solve personality issues. When someone is annoying, you just ship them out to another department. Another great tactic was Tokyo. If you found yourself "promoted" to a prominent position in Tokyo, someone had screwed you and you might as well quit, because you were probably never coming back.

Paul Brooke had been at Morgan Stanley long enough to rub lots of people the wrong way and he found himself promoted to director of research in Tokyo. He was smart enough to know he had been sent to purgatory, and started working on a way to get back to New York. Fortunately for him, the Wellcome Trust had decided to take public part of its private endowment to raise cash for additional research. Brooke immediately flew to London and attached himself to this huge multibillion deal, which was sure to pay Morgan Stanley tens of millions in fees. He'd found his return ticket out of Tokyo.

Paul Brooke told me this story as a cautionary tale. Not one week later, it was announced that John Huller would be the new director of research in the Tokyo office.

• • •

I kept moving up in the *Institutional Investor* polls. Third place last year, second place this year, edging out Tom Kurlak, who had just recently been locked in as number one. It was exhilarating for me, but no one at Morgan Stanley gave a shit. I had a vote. As long as I didn't lose it, I was fine. I did get a promotion to principal. My compensation stayed the same, but two things happened. A piece of my compensation was paid in Morgan Stanley stock and held in escrow until I left the firm. Gee, thanks. And I was allowed to use the executive gym. Making principal at Morgan Stanley was known as the $50,000 towel charge. I went to the gym once and saw strategist and hedge artist Byron Wein sweating like a pig while going one mile per hour on a treadmill. I never went back.

Years of following these chip companies were wearing me down. Chip companies didn't wait for the economy to turn down to go into their own recession. I was strapped into a roller coaster, trying to pitch long-term stories like Intel that went up 2 points down 1, up 2 down 2, up 3 down 1. The direction was up, but the industry was its own Vomit Comet.

Every vacation I had was ruined. My wife and I were at a bed and breakfast in the middle of nowhere in Sweden. At midnight, there was a knock on the door. I told my wife to ignore it. She thought someone had died. I told her I guaranteed that Intel had just blown up, and that I probably needed a good night's sleep. Sure enough, the urgent message in the morning was to call into the morning meeting in New York to defend my Intel recommendation.

Another time we were on a beautiful island in the Caribbean. The phone rang and it was my mother-in-law saying a very rude man named Stanley Morgan was calling and desperately needed to talk to me about an emergency. She sounded frantic. I called in and the emergency was that Larry King had done a segment on the danger of cell phones because of the possibility that they caused brain cancer, and Motorola's stock was plummeting. OK, so that ruined my trip and I've hated Larry King ever since. I was up way too early the next morning calling into the morning meeting saying "back up the truck and load up with this stock if everyone is dumping it."

What finally set me off, though, was an analyst meeting for AMD at the San Jose Fairmont, run by their flamboyant CEO Jerry Sanders. I didn't like the guy, and I don't think he liked me. In the *Los Angeles Times,* I was quoted by a reporter one weekend as saying that Jerry reminded me of a cross between Chuck Barris from the Gong Show and Hugh Hefner from *Playboy.* I spent two hours on the phone with someone from the company saying it was a tongue-in-cheek comment, and I was sorry, even though I really wasn't.

I had just arrived at the Fairmont and was walking in the front door when up pulled this beautiful baby blue Rolls Royce convertible, with the top down. The chauffeur jumped out, ran around to the passenger side and opened the door for none other than Jerry Sanders and his flaming white hair and shit-eating grin. Sanders apparently also had a separate Rolls Royce and driver in LA for his commute from the airport to Malibu. I wanted out right there and then. I was an analyst getting beat up on a roller coaster analyzing each blip in the industry, when most of the money had been made 15 years before by buffoons like Jerry Sanders. I decided I wanted in on the early stage of the next big industry rather than fighting over the scraps of the last one. Not as an analyst or banker, but as an investor. It was a stepping-stone moment.

Barton Biggs had made me technology strategist, although he failed to mention it to anyone else. I took that ball and ran with it, writing strategy pieces trying to identify new markets for investors, and secretly for myself. I started writing about online worlds, interactive media, and video. I even used the word Internet once or twice, but it was way too early for that. In 1992, the Internet was a way for academics to pretend they were doing important things.

It got me in serious trouble. I downgraded many of the smaller chip stocks I no longer felt like following. Frank Quattrone would call and ask me why I had to do that. There was always a reason to stop recommending stocks, but I didn't have too many good reasons. I was just sick of them. To be fair, Frank Quattrone never once told me to recommend a stock. He'd beat me up to try to convince me, but he never said, "You have to recommend a stock or you won't get paid." There were other ways.

In the meantime, Frankie was a fan of the strategy pieces I was writing. He grabbed a pile of them and shoved them under the nose of Rich Karlgaard, the editor of *Upside* magazine, a brash technology and finance magazine in the Valley. The cover of one of the first issues of Upside asked, "Has Silicon Valley Turned Pussy?" with a caricature of Apple CEO John Sculley. Rich called the next day and asked if he could run my pieces as a column. "Sure, why not?" I said. Of course, I was flattered. So, Frankie launched my journalism career, although I'm still not sure whether to thank him or curse him.

• • •

It didn't matter if I was a strategist or a banking analyst—I still needed my *I.I.* vote. So I kept marketing. Another mandatory marketing trip was the Metroliner special, a 30-hour trek through the mid-Atlantic. Philly for lunch, Wilmington, Delaware, for dinner and a stay at the Hotel Dupont, Baltimore in the morning and lunch in D.C., before boarding the shuttle plane back to LaGuardia.

Over time, this grueling trip added a stop in Wayne, Pennsylvania, home to Pilgrim Baxter. Mary Meeker and I were on one of these trips that found us in the office of Gary Pilgrim, whose name was on the door.

"Tell me the Intel story," Gary said to me.

"Well, the stock has already doubled off its bottom, but their new 486 microprocessor . . ."

Gary Pilgrim cut me off. "Perfect. Let me interrupt and tell you what we do around here. We like fundamentals all right, but we only buy stocks that are already going up, that have momentum. Our success is based simply on finding these momentum stocks and buying them in size. Stocks that go up keep going up. Stocks that don't go up keep not going up. Now what was that about Intel?"

This guy was crazy, but apparently this style of investing and these so-called momentum funds were the hottest thing on Wall Street. Pilgrim Baxter had been posting great performance numbers using this method and lots of other funds were becoming momentum funds. It didn't make much sense to me.

At the end of our meeting, I sheepishly asked Gary, "Wouldn't you want to find stocks that are about to go up, rather than ones that are already going up?"

"We don't have the patience for that," he answered.

There it was in a nutshell. Momentum funds needed performance NOW, and couldn't wait around for a stock to work. Instead, they bought stocks that were "proven," even if they had already doubled. This struck me as more like gambling than investing. Blackjack cards or a craps table gets hot and gamblers at the surreal Caesar's Palace flock to it and bet more.

Wall Street as a casino is an overused metaphor, but an apt one. I am always asked what stocks to bet on, what are the odds that something will work, should the investment be doubled-down? I hate the tie-in. Casinos are for losers. Gambling is a sucker's game. Everybody knows the games are rigged against gamblers—the odds are set so the house always wins. Wall Street is about access to capital for great companies. If you do your work, you can find these great companies that do better than the market. Gary Pilgrim was impatient. He and momentum investors, I call them momos, just wanted to find the hot table.

. . .

Every Wall Street firm had a small cap analyst, someone who followed interesting small companies that either fell through the cracks of coverage, or that no legitimate analyst wanted to bother with. Ruvkun did it in techland, but there were plenty of other small companies that needed analysts. Because bankers often raised money for these small companies, someone had to cover them. It was the worst job on Wall Street. You had to put a happy face on what inevitably was a piece of shit company the bankers rang the cash register with.

Morgan Stanley had a small cap analyst named Ram Kapoor who was known as Ram Ka-make-me-poor. His offense? A banking client named Safeguard Data Systems that blew up and spread shrapnel throughout the Morgan Stanley system. An analyst named Keith Mullins replaced Kapoor. Keith was a little more cautious about touting the more blatant banking clients, but the only reason he had a job was to tout banking clients. Go figure.

I would have lunch with Keith every so often, and noticed he was gone for a few days.

"Been traveling?" I asked.

"No, but I have been to hell and back," Keith answered.

"What do you mean?"

"Andy, have you ever been subpoenaed or had to testify at a shareholders suit?"

"No," I answered.

"Well I have, a few too many times. Just spent this week in court. It sucks. Don't ever get stuck in this position. I was minding my own business, being a good analyst, recommending the stock of some crappy IPO our bankers brought in. It did OK, then blew up and sank like the Titanic, taking tons of people down with it. The lawyers came in and took every piece of paper and all the notes I had on the company. Then I had to testify, and they make you into a complete idiot."

"How?" I asked.

"Like this. 'Mr. Mullins, you work for Morgan Stanley?'

'Yes.'

'It is the top firm on Wall Street?'

'One of them.'

'And, Mr. Mullins, you have a degree from a top university?'

'Yes.'

'And, Mr. Mullins, you made zillions of dollars last year?' The guy had my comp down to the penny.

'Yes.'

'Now, Mr. Mullins, you are a very smart guy?'

'Well, I don't know . . .'

'Mr. Mullins, you are from a top school, work at the top firm on Wall Street and make zillions. You must be a smart guy.'

'OK, I suppose I am.'

'And you recommended this stock?'

'Yes, but . . .'

'And management told you everything was doing well?'

'Yes, but . . .'

'And you couldn't figure out something was wrong?'

Then he turns to the jury.

'Ladies and gentlemen, here is one of the smartest and best paid members of Wall Street's elite who couldn't see something was wrong and was fooled by the management of this company. How do you expect my poor clients to fare when led to slaughter by this corrupt management team? It is clearly the company's fault my clients lost money and I think they should pay for their chicanery and tomfoolery.'"

That was all I needed to hear. I recommended stocks that went down all the time. Not on purpose, of course. Dennis Callahan, the insurance analyst at Paine Webber had it right: never recommend a stock that could go down to individual investors. I went back to my office and emptied my files into the garbage. Not that I had anything to hide, I just didn't want to be a Mr. Mullins to anybody. From then on, I kept only annual reports and product brochures. I would take notes at company meetings and conference calls, which I used to write up my reports. Then I tossed the notes. I never referred back to them anyway.

Sure enough, nine months or a year later, I got a call from the legal department of Morgan Stanley. They wanted to come down and pick up my files on Intel, which had been subpoenaed in a class action lawsuit against that company. I complied, with annual reports, SEC filings and a 200-page bound copy of the technical specifications of the new Pentium microprocessor that some poor paralegal was going to have to make copies of, for no good reason. Every six or nine months, lawyers from some Motorola or Intel lawsuit would subpoena my files. Being the ax in a stock had its downside, but thanks to Keith Mullins, I never sat in the witness stand.

. . .

I went to the Far East once a year to check on chip competition and visit financial clients. The government of Singapore was a big client of the firm, as were several institutions in Hong Kong and Tokyo. I went to see them all. I always came home with a bruised forehead. It took a few trips to remember to duck my head when using urinals in Japan, as they were set up for people five foot six inches or shorter, and I always smacked my head on some overhang above them.

I was in Tokyo on one visit and who should be there at the same time but Ed Greenberg and Dan Reingold. They were calling on Asian telcos about privatization and then visiting clients just as I was. I started to ask Dan Reingold if he was bruising his forehead, but remembered Jack Grubman's "shop in the children's department" line and dropped the subject.

Giving presentations in Japan was interesting. There was usually a translator and you had to pace your presentation to allow the translator to keep up. As I had plenty of jokes in my presentations, the people who understood English would laugh each time I told one. A few moments later, the translator would get to the punch line and those in the room who spoke only Japanese would laugh.

My counterpart in Tokyo, Takatoshi Yamamoto, pulled me aside and asked me about John Huller. (I apologize for the bad Japenglish translation.)

"What you know bout John Hulla?"

"Why?" I asked.

"I no get John Hulla. He wok round, want know what we do. We no tell him, he too nosy. Nobody like him."

"How do you know that?" I asked.

"Watch morning meeting. Translator translate everybody, but no translate John Hulla. He too stupid, he no figure out."

I got up and said a few words, with the translator keeping up. Dan Reingold gave a short pitch, with translation. John Huller said a few words about the U.S. market and deals to work on and there was silence—no translation. I didn't think John Huller had a return ticket to the U.S. Eventually he did make his way back to the U.S., as the head of sales for the boutique Hambrecht & Quist.

• • •

The only thing more annoying than the line "Welcome aboard" when someone started at Morgan Stanley was the line I got just about every day when I left to go home. Someone at the elevator bank would inevitably be getting off or walking by at 5 p.m., 6 p.m., even 7 p.m. just as I was leaving and say, "Half day?"

"Yeah, yeah." I'd say, or "Client dinner. Al Gordon, Marcus Schloss," or "Gonna do some stuff at home." Finally, I got tired of the half-day comment and started replying, "It only takes me a quarter day, what's your problem?"

Mary Meeker was one of those stay-all-night analysts. I'm not sure why. Maybe the most creative thoughts came late at night. Maybe she was just terribly unproductive. Who knows? But it started to bother me. 10 p.m. phone calls asking for advice were easy to deflect, as I was asleep. But just about every Saturday morning or afternoon, the phone would ring. My wife would say, "Mary has a short question for you," roll her eyes and hand me the phone for what was a several-hour-long tutoring or therapy session about how to deal with a rabid sales force. I walked her through it, but we eventually stopped answering the phone at home. Ever.

• • •

Every year, Barton Biggs would host an exclusive retreat for the top portfolio managers in the world at Lyford Cay in the

Bahamas. Sir John Templeton, the highly successful mutual fund manager, lived there and attended the conference. Barton's idea was to get them all in the same room, and have them discuss how they saw the world and the market. He would write up what people were thinking in general. Unremarkably, almost everybody thought the same thing. Group think. The issues or market view might vary each year, but the group mainly agreed. Of course, this is valuable information, because it sets the consensus, and helps the really smart ones think differently from every one else.

"I am the firm's technology strategist, so I should go to this event too, right?" I asked Byron Wein.

"Dream on, pal," was more or less the feedback I got. "This is very exclusive, and no place for analysts."

"But, but, but, I *am* a strategist," I said, trying not to sound as though I was begging.

"Yeah, I had heard something along those lines, but I never got the memo."

So I went to Barton Biggs and asked him if I could go.

"Yeah sure, why not?"

I booked the trip through Barton's secretary, and flew down with Steve Roach, a regular at the event. When we pulled into the resort, I got a pretty funny look from Byron, a "what the hell are you doing here?" look. It was a great moment.

I came prepared. Knowing I would get about ten minutes to say something interesting to this amazing group, I scrambled my brains for weeks, and finally came up with an idea. I was bullish on technology, but this was 1992, and few other people even cared. Most portfolios were under-weighted in technology, meaning they owned less than the percentage that technology stocks represented in the market. Most of these guys just owned IBM because new stuff scared these old dogs.

My pitch was real simple. "Overweight technology." It was met with a lot of groans. "OK," I continued, "I knew I would get that reaction. Even if you don't change your weighting at all, buy a synthetic IBM."

"A what?" someone asked.

"IBM is dead, a dinosaur about to collapse. For every three shares of IBM, buy one share each of Intel, Microsoft, Motorola, Adobe, and Novell. Big changes are going on inside the sector." The reaction from the room was ho-hum.

In retrospect, Cisco might have been a better choice than Novell, but it didn't matter. Over the next few years, even over the next decade, the synthetic IBM completely obliterated the real IBM, as companies and as stocks.

. . .

The cool thing about being a strategist was that no one inside Morgan Stanley knew what I really did. This pissed off Frank Quattrone, as he wanted to tell me what to work on—how dare I decide for myself. But I did.

The media bankers loved the stuff I was working on, and dragged me to see their clients. I was stepping on toes because Morgan Stanley had a media analyst, Alan Kassan. He was OK with what I was doing, when I explained it to him. The firm's media bankers (not Frankie) took me in to see Disney, CBS, Paramount, and TimeWarner. Sumner Redstone of Viacom invited me to speak at an offsite seminar meeting for his company down at the Breakers in Florida. I met with Nintendo and Sony and Sharp in Japan. Rupert Murdoch and Barry Diller had read my pieces. John Sculley wanted to meet me and talk about interactive media, so into Apple I went. I don't think he remembered me from that "no place for Steve Jobs" analyst meeting years ago. Rob Hersov, a media banker in Europe, took me to see all his clients, from Carleton to Pearson to Kirsch. This sure beat the hell out of tracking DRAM memory prices, let alone digging ditches.

Frank Quattrone called and told me "We're not going to make a penny off any of this multimedia bullshit."

"Yeah, maybe," I said.

"Andy, these semiconductor companies are cash-guzzling machines. They need to raise money ALL the time. And we are just the guys to do it for them."

"Got it, Frankie," I said.

"You know, Andy, I just don't understand. We made George Kelly into the analyst he is today, and are about to make Mary Meeker into a top analyst. Why don't you let us do that for you?"

"Got it, Frankie," I said. I think I said "asshole" after I hung up the phone.

I had "made" myself, or at least I thought I had. I was a top-ranked *I.I.* analyst, but had not become a lap dog to bankers. Did I have that much integrity? Nah, I could be swayed by money. I had figured out long ago you work on Wall Street to get paid. But my clients were institutions for which I felt an obligation, to make smart and right stock calls. All you have on Wall Street is your reputation. If you lose that, you are toast.

I just couldn't bring myself to recommend stocks that I didn't think would go up to my clients. Maybe if I hadn't started out at a non-banking firm, and hadn't had Bob Cornell beat into my brain the game of being right first, I could have been a lap dog. Who knows?

I got called into a meeting with Moe-Larry and Jack Curley. "I just got off the phone with Frank Quattrone," Curley launched, "and he was complaining about you."

"Really?" I asked.

"Something about not making a penny off your multimedia bullshit research. I'm not sure what he is even talking about. Are you working on something besides semiconductors? Let me put it this way. You don't want that type of feedback from bankers when it comes to bonus time around here. Understand?"

"Loud and clear."

Frank was right, for a while. We didn't make a penny from multimedia, but I was having fun. I was on panels with Steve Case, CEO of a small public company called America Online. I spent time talking strategy with Jim Clark, Chairman of Silicon Graphics. I got to know Rob Glaser, who was in charge of multimedia at Microsoft.

I got a call from Skip Paul, one of the senior executives at Universal Studios. "Next time you're in LA, stop in and talk." Universal was owned at that point by Matsushita, and Skip not only spoke Japanese, he had worked in the video game industry and was the point person on technology with the studio. I booked a trip to Los Angeles the next week.

Skip loved talking strategy, and thought the whole world was going to change around online worlds, and that studios were sadly going to be left behind. It was lunchtime, and when Skip asked if I'd like to go to the commissary for lunch, I could only think of all the Johnny Carson gags about the NBC commissary. "Sure, I have time."

I wasn't disappointed. There were lots of people in costumes, probably movie stars, but I didn't know, or care. On our way to the exclusive tables at the back of the room, Skip introduced me to two nice old men, one with Coke-bottle glasses. The men were Lew Wasserman and Sid Sheinberg, who despite selling Universal to Matsushita, still ran the studio. Sid Sheinberg said to me, "I read your pieces. I'm still confused."

Lew Wassserman jumped in, "Doesn't Morgan Stanley still have that Alan Cay-san?"

"Yes, Alan Kassan is our media analyst," I said.

"Well, his stuff sucks," Lew said.

Skip Paul yanked me away saying, "OK then, let's go eat."

. . .

When I got back to New York, Frank Quattrone called. "I can't even believe it, but John Sculley wants to see you and me in his office next week. I've invited Mary Meeker and Bill Brady to the meeting, too. Let's work up something unique for him, he could do lots of business."

The four of us brainstormed. I was in print with the synthetic IBM stuff, so perhaps we could use something like that. We decided we would pitch Apple on putting together a synthetic Microsoft. Apple's operating system, Borland's programming tools, Lotus applications, Adobe desktop publishing software and other companies who worked with video and online stuff. Maybe we even had them buying America Online. On and on until the whole thing was put together. I could hear Bill Brady sigh on one of the conference calls. He was going to have to pull a few all-nighters putting together the book on this one.

. . .

We piled into Apple the next week. Sculley was there. Their CFO Joe Graziano was there too, along with their general counsel Al Eisenstat. We handed out the pitch books and jumped into it. Apple needs to put together a synthetic Microsoft if they are going to win the coming interactive media wars. The Apple guys listened intently, nodding occasionally. I was excited. I was proving in real time that there was money to be made from this "multimedia bullshit."

We finished our presentation, mentally exhausted. Preparing to speak, John Sculley pushed his chair back a bit. "That was great, thank you. But I have a question for you. Why doesn't AT&T buy us? I mean, they are in the computer business but they have this unfair advantage of using ratepayer money to subsidize it."

Sculley then launched into a tirade. "You can't believe how hard this business is. We enter every quarter having no idea how many boxes we are going to sell. We cut prices every quarter and hope we can sell more of these things to grow the business. It's hell. AT&T ought to buy us, they would have an easier time of it."

Frank politely said he would check with AT&T, and we filed out. In the lobby, Frank growled at me, "Well THAT was a waste of time."

I just shrugged.

In the meantime, I was meeting with a lot of interesting small companies. Game companies, online tool companies, networking companies, video chip companies. They all said they liked my views, but they needed investors, not investment bankers. I started thinking that I was in the wrong business.

. . .

Mary Meeker started doing better with the sales force. Her calls got better, and some stocks she recommended actually went up. She got good at being an analyst, but I'm not sure she ever got comfortable with institutional clients. They could ask her about banking clients, since she was involved in doing due diligence with the bankers. Frank Quattrone was almost anal about doing thorough due diligence on companies since he had been asked to testify in the Miniscribe law suit. Nothing that had happened at Miniscribe was his fault. The company faked the numbers. But, Frankie said never again, and beat up his bankers to dig deep to find problems at companies and then run away at the hint of any improprieties.

Mary Meeker had an analytical crew she could count on to wade through due diligence and analysis of banking clients, but I don't think she ever felt comfortable with her own analysis. The bankers were her filters. Bad move.

• • •

Just after the *Institutional Investor* poll had come out, I stopped into Mary Meeker's office with some advice about how it was all a marketing game and how she could use the sales force to her advantage. Her head barely came above her desk and she was at a strange angle, sort of kneeling on her chair. Chair is a bit of a stretch. It was one of those weird contraptions with padding in strange places that you sort of mounted and it took the pressure off of your back and perhaps your brain.

"Hey Andy, you going to Comdex this year?" she asked. Comdex is the big personal computer industry trade show in Las Vegas every November.

"Yeah, I'll be there." I had a bunch of interesting small companies to see.

"Uh, well, I've got about a million important meetings all around Las Vegas, and there is no way I can make them all in time, so I asked Jay Cushman if it was OK to hire a limo."

"A limo. You're kidding, right?" I asked.

"He agreed, but only if the rest of the tech team going to Comdex use it as well, so that means you. Will you use it?"

"Sure." Of course, I preferred to cut cab lines like I do every year—a Comdex hobby of mine—but a limo would make things easier.

Although I ran into Mary at a few big meetings at Comdex, I never did set foot in the limo. I did see it once, a nice black stretch with Mary ensconced in the back, like the Queen.

The week after Comdex, Mary came into my office, just as I was settling in with my back to the door, feet up and a trade magazine propped open. She was interrupting naptime.

"Can I ask you something?" she asked.

"Sure," I said.

"Uh, that limo?"

"Yeah, I got some great usage out of that," I said in my best snide tone.

"Oh yeah, sorry. My question: The bill is $3000 and I told Jay it would be around $500. What do I do?" she asked.

"Al Gordon, Marcus Schloss," I immediately answered.

"What?" she asked.

"Al Gordon, Marcus Schloss."

"What does that even mean? Who is that?" she asked.

"No idea. The guy calls me once a quarter and asks me for my earnings estimates. I've never met him. However, I happen to wine and dine him constantly. We've been to the finest restaurants in New York, and take lots of cab rides together around town. I'm sure I've dropped at least three or four grand of mother Morgan's money on this guy over time. Should I go on?"

"Nope, I think I get it." And Mary did get the secret to better stock analysis: creative expense accounting.

• • •

Morgan Stanley was raising money for Silicon Graphics, which constantly went to the well for money for new buildings and expanded sales offices, the perfect banking client. I figured I would check it out, eat some rubber chicken, and maybe learn something about what they were doing with 3D graphics for interactive media. I sat down next to Frank Quattrone, who introduced me to the guy sitting on his other side. He was Bob Harris, a banker from one of the other underwriters.

"Oh, you're the clown writing those goofy multimedia pieces," Bob said.

"That's me," I said.

"Actually, they're pretty good, everyone's talking about them. You're wasting your time at Morgan Stanley."

I had no idea what he was talking about, but called him up a few weeks later to find out what he meant. I went into his office in San Francisco, which, like Frankie's, was filled with tombstones from deals he had done. This time I paid attention to the names. Microsoft, Sun, Silicon Graphics, Oracle, and on and on—a who's who of Silicon Valley.

"This is a pretty impressive group of companies," I said.

"Yeah. But Andy, they were just a bunch of shit-bag companies when I first met them."

I immediately thought to myself, hey, I know a lot of shit-bag companies, too. Maybe they can turn into a similar who's who list. I was pretty delusional.

Andrew Wallach, my lunch partner on my first day on Wall Street, was right. Research was just a stepping-stone to something else. I just hoped that something else was another stone, not a lily pad. I worked on taking that step.

• • •

In the meantime, I began kissing frogs. Bill Brady and I went to see a company, Mediavision, that made sound cards and was one of the first multimedia plays. I was interested in the company, but something was wrong with them. There were two top guys, neither of whom gave off good vibes. Frank Quattrone turned it down, before we got any further. Good thing, because the company eventually went public and apparently shipped empty boxes to make their quarter and the whole thing blew up in a huge disaster.

We dodged a bullet, but then my phone started ringing. Avid, the leading video editing company out of Tewkesbury, Massachusetts, wanted us to take them public. 3DO, a secretive video game platform company run by Trip Hawkins and backed by Kleiner Perkins, AT&T and TimeWarner, wanted to go public and picked us. They were my deals, but Frankie got Mary Meeker involved with both of them, which annoyed me a bit, but then I stopped caring.

The deals were great for me, but unfortunately, a little too late. I had already agreed to move out to the San Francisco area and work with my own shit-bag companies as a partner with Bob Harris at Unterberg Harris. I was hoping to become a venture capitalist, but would be an analyst and maybe even an investment banker for a while in order to pay the bills. You could do this at a small firm. If anyone was going to use me as a piece of Wall Street Meat, it would be me.

I had seen the writing on the wall. Frank Quattrone and, to be fair, the whole firm, were more interested in banking analysts like Mary Meeker and George Kelly than in stock picking analysts. I was getting dinged for not bringing in more deals, until Avid and 3DO came along, albeit too late.

The timing was awkward, because I had to wait for my final bonus to be paid at Morgan Stanley before resigning, otherwise, it would disappear in a puff of smoke. So I worked on Avid, helped sell the deal, and even went on some of the road show. They came to New York for a big presentation, and I joined management at a dinner at Ben Benson's (charge it to the deal). It was strange to start and end my traditional Wall Street career at the same restaurant. The CEO of Avid asked me why I was carrying two big bags with me. "I often take my work home," I answered. In reality, I had all of my computer files and Rolodexes jammed into the bags. I was resigning the next day.

What a day it was. No sooner had I submitted my one-sentence resignation, than Frank Quattrone called to tell me Morgan Stanley had just received the mandate to lead a huge warrants deal for Intel. "Congratulations," Frank said, "your research is really paying off."

"Hey Frank, this is great news, but I just resigned about five minutes ago," I told him.

"What? You can't leave," Frank screamed.

"Yes, I can," I said calmly.

"Well you have to stay, at least for this last deal. What if we made you an offer you can't refuse?" Frank asked.

"Frank, what are you going to do? Put a horse head in my bed?" I asked.

"Maybe," he chuckled.

When news got around that I was leaving, I got tons of calls from friends inside Morgan Stanley.

"Good move."

"I can't wait to get out of here, too."

"You obviously quit because they didn't make you Managing Director."

"Take Mary with you."

The Intel warrants deal got done without me, and apparently, without Morgan Stanley breaking a sweat. Avid went public and became a successful company. 3DO went public with a bang and faded with a whimper when no one bought their boxes. And I was almost done being a Wall Street analyst. It would take several years of detox to get it out of my system. My friends Jack Grubman and Frank Quattrone and Mary Meeker would all become superstars. At the end of the day, I was glad not to be one of them.

CHAPTER 10

Netscape IPO

I futzed around for a few years. I did some deals by pretending to be an analyst. Some great ones like chip company C-Cube Microsystem's IPO as well as raising money for game company Activision. I also did some pieces of shit like Interfilm that did interactive systems for movie theaters, an instant dud. In the background, I raised a small venture fund with a funky name, the Interactive Media fund. I began chasing down my own, as Bob Harris would put it, shit-bag companies.

I hid out in San Francisco, away from the rat race of Wall Street in New York. Bob Harris did a lot of work with Silicon Graphics, with me as an observer. I was spending time with Jim Clark who was desperate to get out of Silicon Graphics and do something more interesting.

I did keep tabs on everyone. Rod Berens came out of retirement and took over capital markets at Salomon Brothers. One of his first hires was Jack Grubman, out of Paine Webber. That gave me a good chuckle.

Mary Meeker kept cranking out her emotional tirades. She was still scouring the Valley with Frank Quattrone looking for deals. They did a few, nothing really interesting. Morgan Stanley had hired Alan Rieper, the Grim Reaper, to replace me. I'd run into Frank every once in a while and he would just shake his head and tell me I was wasting my time and should come back and be a "real analyst" again. He also voiced his frustration with the powers-that-be in New York. He had this incredibly successful "boutique within a bulge bracket firm," he owned Silicon Valley, but he couldn't hire more bankers, couldn't influence their bonuses, couldn't hire the analysts he needed to bring in more deals, couldn't get the sales force to focus on his deals vs. paper companies or all the other industrial crap that Morgan Stanley was de-leveraging. I must say, he was right, but I didn't really feel his pain. I couldn't have cared less. I was glad to be out of the analyst game and working on my stepping-stone.

Jim Clark found his new "more interesting thing" to do. Originally called Mosaic Communications, Netscape would spark a new era.

One day in August 1995, the ballroom at the Waldorf Astoria quickly filled up. Most of Wall Street was on vacation, at the Hamptons, telling tales of great trades and huge commissions. But there was a hot IPO in town, and many were curious to hear the story of this quirky Internet startup, Netscape. If you looked closely before the presentations started, you would have seen a guy in an ill-fitting pin-stripe suit, a cheap Casio watch and a mustache in need of trimming. Frankie Q. scanned the room, knowing that this deal was a defining event for his technology group. Arms folded, with a Mona Lisa smirk on his face, he would greet the occasional institutional investor with a quick hello. He knew the only reason they were paying homage is that they hoped to get an allocation of shares in what was clearly going to be a hot, hot, hot IPO, sure to pop soon after trading began. Near Frank, you would have also observed Mary Meeker, chatting with a small clique of clients who were badgering her for earning estimates, growth prospects and, I suppose, how she really felt about the company.

The Netscape IPO scared the hell out of me. Not because it made Jim Clark a billionaire, not because I had had a shot at being involved (man, I could write a book about that), and not because Frank Quattrone and Mary Meeker were involved. It scared me because it was worth so damn much. The market was valuing Netscape at $2.2 billion. This for a brand-new software company that had never really been tested. Bob Cornell had taught me about duration years before. Netscape's duration must have been five years instead of the "normal" 18 months. The duration would stay this long for tech stocks for the rest of the decade, allowing lots of companies with high growth and no earnings to hit the public markets.

And then what really annoyed me was that Netscape's IPO was a hundred fold oversubscribed, breaking the five-year-old record of Xilinx for the hottest IPO at Morgan Stanley. How dare they?

The boom was starting.

. . .

Business was there for the taking. In the post-Netscape IPO boom, quality was never really an issue. Investment bankers used to insist on two consecutive quarters of profits before taking a company public. Now suddenly, it was all about growth—How quickly can you grow your sales? If you are doubling, we'd be happy to take you public.

After years of living in Frank Quattrone's wake, Goldman Sachs finally woke up. It was the old "late night at a bar" syndrome. If the only women who will talk to you are below your standards, lower your standards.

Boutiques were doing Internet deals, and the institutional ducks were quacking. At every investor conference I went to, I would be stopped in the hall by old clients with the same question, "How do you play this Internet thing?" Others just bought at any price. Tech fund Amerindo quickly built a million shares position in Netscape. Their philosophy was "let the winners run." Not wanting to be left behind, more growth funds converted into momentum funds, momos.

In April 1996, the ducks screamed. Lycos, a search engine company inside CMGI, went public at $16 and traded up to $22. Boutiques H&Q and Alex Brown were the underwriters. Two days later, boutique Robertson Stephens priced the IPO shares of another Internet search engine company, Excite. I made the drive up to San Francisco to the Excite road show presentation and it was pretty ho-hum. The CEO used to make documentaries for National Geographic. The company was losing money. It didn't matter. Shares were filed at $12–14, priced at $17 and closed the first day at $20.

Boutiques could get away with taking these early-stage companies public, despite them being barely a year old. They were growing like weeds by losing money. If the stocks blew up, well, so what? H&Q or Robertson's reputation wouldn't be tarnished. It would probably be enhanced as an underwriter that is willing to try new things. Some stocks work, some don't, but they are worth a try.

But Goldman Sachs, who should have known better, couldn't resist. A week later, they priced shares of yet another Internet search engine company, Yahoo. They had sales of $1.4 million in their short ten months of business. They had lost $643,000. But Yahoo's growth was not just like a weed, more like a weed on Miracle-Gro fertilizer. CEO Tim Koogle laid out a media strategy on their road show. Venture capital firm Sequoia Capital, run by Don Valentine (who backed Apple and Cisco), gave it an air of legitimacy. Goldman Sachs took the plunge. At the time, it was the smartest thing they did. The ducks were overdosing on amphetamines. The stock shot up from $24.50 to $43, valuing the company at $1 billion.

• • •

This sent Frank Quattrone over the top. He had finally had enough. He had the hottest IPO ever at Morgan Stanley, or anywhere on Wall Street, and the big wigs in New York still wouldn't give him control of anything. Netscape was still small potatoes compared to the blue chip companies Morgan Stanley dealt with. It was the Bob Metzler view: "Is it alive? Good enough." Besides, Morgan Stanley really believed that Morgan Stanley brought in the deals, not individual bankers. It was a meritocracy, not a star system. If I were Frank, I'd be pissed off too.

Goldman Sachs was doing Internet deals. Morgan Stanley's "boutique within a bulge bracket" firm was really just bullshit. Frank's mentor, Carter McClelland, had recently gone to Deutsche Bank to help them create a technology presence. Carter's job was easy. He hired Frank.

Carter let him create a real boutique, where Frank was in control of everything. In this groundbreaking deal, at least the way I heard it described, Frank and his group got a percentage of revenues for their bonus pool. Secretly, Frank negotiated a put option with Deutsche Bank. A few years out, he could sell the entire group back to the Bank at some multiple of their revenues. This gave Frank a huge incentive to do deals, any and all deals that generated revenue. Quality was not Job One, as they say at Ford.

Along with Frank went Bill Brady, his old associate, and George Boutros, one of the best mergers and acquisition bankers on Wall Street. The three of them flew to New York, set up in a hotel room. I suspect there was some standoff-clause in their departure agreement with Morgan Stanley that said they couldn't recruit anyone to work for them, but, as word spread, people found them and interviewed to be part of the Deutsche Morgan Grenfell Technology group.

Notice that this was not the "Technology Banking group," or "Technology Research group," or "Technology Trading group." Instead, the umbrella was held high over *all* of them. They hired more that 150 people within a week, to work in the old EF Hutton building in New York, or in Menlo Park, California, so as to be near venture capitalists who were directing the IPOs of their investments. Frankie had analysts, bankers and traders—everything that was needed to create that "boutique within a bulge bracket" firm.

• • •

While all this was going on in 1996, Fred Kittler, my old client from JP Morgan, and I started our very own firm, Velocity

Capital Management. Fred insisted the name start with a V, something about being Pynchon-esque. We would raise a fund to invest in both small public companies and later-stage private companies. How hard could that be?

Hard? It was a royal pain in the ass. We struggled to raise money. We started with a whopping $10 million in assets, 10% of our goal, but trudged ahead. The stepping-stone was the size of a pebble. We figured that as we proved ourselves, capital would show up. I was finally on the buy-side. Wanting desperately to prove myself, I was keenly interested in how the sell-side could help and how I could work with them. It wasn't very encouraging.

I ran into Frank at some event or another, and he breathlessly walked me through what he was building. He didn't recruit me, but he pitched me as a client of his new boutique. He did try to hire George Kelly, but George was too close to retirement to start something new. Frank really lamented when he talked about Mary Meeker, whom he tried hard to hire. She wouldn't budge. Morgan Stanley finally figured out they were being raided and set up retention plans for analysts and others they wanted to keep. The star system crept into staid old Morgan Stanley, at least in research (the new West Coast banker was someone out of syndicate). Mary, as Netscape's analyst, was a keeper. Frank mentioned something about how he was going to have to invent her if he couldn't hire her. There was business to be done, banking fees to generate.

CHAPTER 11

Quacking Ducks

Gary Pilgrim must have been spreading apple seeds throughout the mutual fund industry.

Back in 1980, a scant $40 billion was being professionally managed by mutual funds. The rest was in pension funds, foundations like the Ford Foundation, or in banks. These "professionals" were terrible, most couldn't pick stocks that did better than the market. Books were being written about chimpanzees and dart throwers doing better than professional money managers. John Bogle at Vanguard in Pennsylvania had a solution, i.e., if you can't beat the market, just become the market. Index the whole thing. The bulk of the market was represented by the Standards and Poor (S&P) 500 index, the top 500 valuable public companies in the U.S. Bogle offered an index fund that did neither better, nor worse than the market, and it caught on as a savior of investors. Money came out of banks and into mutual funds, much of which were index funds. By 1996, over $1 trillion was in mutual funds.

But indexing is dull. Those looking for better returns tried other ways to beat the market. General Electric, worth $240 billion makes up 4% of the $8 trillion S&P 500 index. If it halves in value one day, but two $60 billion valued companies double in value that same day, the index doesn't budge. Ho hum. But someone can make money as the innards churn. Gary Pilgrim had a formula: just wait for stocks to go up, prove that they can go up, then buy them and ride them for all they're worth. The cowboy saying is "ride-'em hard and put-'em away wet." It's that "put 'em away wet" part that no one has quite figured out. The hardest call for an investor is to sell when the going is good.

As the bull market picked up steam, lots of funds became momentum funds. Eventually, almost every growth fund became a momentum fund, whether they admitted it or not. Janus out of Denver was a classic momo—they even had a Janus Twenty fund, narrowly focusing on twenty momentum stocks. These momos were hungry for ideas that worked, and became the ducks that were quacking and buying deals from the Street.

Goldman Sachs had discovered a great one with Yahoo. Despite lowering their standards, the damn thing worked. It was worth $1 billion on its first day of trading. I thought it was a bargain at half the price. Shows you what I know, it kept going up.

The problem was that there just weren't that many Internet companies to go around. Every momo had to own a piece of Yahoo, and bid it up. As the performance of momo funds improved, more money came out of index funds and into momo funds. What was needed was more public Internet companies.

Not a problem, for there were plenty of companies that

could quickly adopt the Internet, put dotcom at the end of their name, and get fed to the ducks. Wall Street was happy to oblige, for a modest 7% fee.

. . .

One of the first high profile deals Frank won was Amazon.com, an online bookseller in Seattle. Morgan Stanley and Mary Meeker put on the big push to get the deal, which, like Netscape, was backed by Kleiner Perkins and venture capitalist John Doerr, who was as hot as a pistol.

Jeff Bezos, Amazon's CEO, had an infectious spirit of doing something yourself the right way. He left New York for points west to pursue his dream. Perhaps he saw the same thing in Frankie. Plus, Frank was early in putting Bezos in front of investors, prominently promoting Amazon at his Technology Group's conference in December of 1996. Frank Quattrone and Deutsche Bank won the business.

Before congratulating Frankie too much, however, it turned out that Morgan Stanley had this tiny little problem. They owned a piece of Barnes and Noble Booksellers in one of their merchant banking funds. Morgan was stuck with the conflict of an old "bricks and mortar" company just as "online" was taking off. They eventually bought retail broker Dean Witter to compound the same problem. Some people never learn.

Not getting Amazon was a blow to Morgan Stanley. Mary was telling people that Morgan Stanley really didn't want to do the deal, as Amazon was probably not ready. I ran that past Frank when I saw him next and he said, "That's funny, when I ran into Mary and those bankers in Amazon's lobby as we finished our pitch, it sure seemed like they wanted to do the deal."

Amazon was ready to price in May 1997. The ducks were so hungry, and demand was so strong, that Deutsche Bank increased the number of shares from 2.5 million to 3 million and set the price at $18 versus the original $12–14 when it had filed the deal. It hit $29 and closed near $25 on its first day. Quack-quack.

• • •

Morgan Stanley may have lost the business, but Goldman Sachs smelled opportunity. They converted a salesman, Michael Parekh, who knew how to open a browser on his PC, into an Internet analyst, and he started hunting down deals. Corporate profits were strictly optional. Companies just had to show a propensity to grow. Netscape and Yahoo and Amazon were the models for future deals to be done.

Fortunately, our fund was an investor in a few of the companies that Goldman had found. Real Networks, run by former Microsoft multimedia head Rob Glaser, was brought public by Goldman in November 1997. Great company, probably a little early for their IPO but there were no complaints from me or the other venture capitalists or the ducks that send the stock up on its first day.

One of the investments from my old Interactive Media venture fund was Exodus. We had invested in the first round, valuing the company at no more than $7 million. There wasn't much to the company, just an optical data line connected to a few servers in a dilapidated office park in Santa Clara. They ended up raising a ton of money, and had built huge air-conditioned web hosting data centers, where Yahoo, Excite, and Lycos and just about every other web company would put their servers. In March 1998, Goldman priced the IPO of Exodus Communications, a web hosting company that was bleeding cash and absolutely needed to go public. Goldman got the deal despite Frank Quattrone having raised the last private round for Exodus. It traded up, and eventually was worth billions. But then again, what wasn't?

A few months after their IPO, Exodus held an analyst meeting at their brand new data center. What an absolute joy for me to attend a meeting as an analyst who was already an investor in the company from its inception. This had been my goal since sneering at Jerry Sanders and his chauffeured Rolls Royce. Of course, I still had a long way to go—I drove to the meeting in my Jeep.

After the formal meeting, there was a tour of the data center, starting at the guard's station at the entrance with three-inch-thick bulletproof glass. This place was like Fort Knox—they pitched it as impenetrable. Inside, there were metal cages filled with servers. Looking around, you could see servers running Yahoo Mail, Inktomi's search engine, and Excite's message boards. If anything happened to this building, half the Internet would be taken down.

The tour ended up in a small room off to the side of the building. In it were bundles of cables connected to Cisco routers. The tour guide explained that this was where the fiber optic lines from Pac Bell, Worldcom, and AT&T came in to the data center delivering bandwidth, which the router then dispersed to the various services. The room cleared out, leaving an analyst from one of the boutiques and me. We poked around for a while, looking at the equipment being used. I sidled up to the analyst and in a half-whispered voice asked,

"You want to be on the front page of the *Wall Street Journal* tomorrow morning?"

Who wouldn't? But, not sure what I meant, the analyst asked, "Uh, what are you talking about?"

"Just yank out that orange fiber cable over there, and half the Internet will disappear. You'll make the front page for sure."

Shooting me a nasty look, the analyst ran out of the room. Too bad, those marketing opportunities are rare. He could have been a star.

At the end of 1997, we received a call from Inktomi saying they were doing a round of financing. They were in the search engine and web caching business but didn't have any name brand customers. Goldman wouldn't take them public unless they got Yahoo as a search engine customer or AOL as a cache customer. Fair enough. It looked interesting to us, and we figured that within a few years both of these would be big businesses, whether or not Yahoo or AOL were customers. We invested and within six weeks they closed deals with both Yahoo and AOL. Goldman teed them up and they were public by June. The remaining shreds of standards at Goldman Sachs opened a window, albeit a tiny one, allowing us to invest in Inktomi. Goldman would quickly lower their standards to almost nothing.

All three of these companies could have used another year or three of gestation to get their business model right and customers established. Had that happened, they would have been much stronger companies today. Only Real Networks is still around.

• • •

It's an almost Pavlovian response. Bring deal—trade up. Goldman brought out Broadcast.com in July 1998. It was an interesting company. I had met founders Mark Cuban and Todd Wagner back when it was called Audionet. They used Real Networks software, and at the same time that they competed against them. This didn't stop Goldman, whose attitude was why worry about competitors. So, they did the deal. It popped 250% the first day.

Soon though, Goldman's deals started to really stink. Goldman Sachs filed the deal for GeoCities. They had a huge series of sites on the web, but almost no business. They generated about $1 per subscriber, while America Online generated $55–60 per subscriber and the *Wall Street Journal* over $600 per subscriber. This thing was a serious piece of skank. Surely Goldman Sachs could see that. No way. The ducks quacked and it traded straight up, and was eventually sold to Yahoo, as was Broadcast.com.

• • •

Frank Quattrone ran with Amazon and lots of tech deals. At Deutsche Bank, he was growing his business, hiring more analysts and bankers and branching out to all the nooks and crannies of the tech business that he hadn't been able to do at Morgan Stanley. I think he was really having fun.

As Goldman was ramping up, Frank ran into a small problem. The way the story was told to me, someone at Deutsche Bank headquarters in Germany finally got around to reading the contract and agreement with Frank and the Technology Group and was shocked to learn about a put option. Frankie had the right to sell the entire Technology Group back to Deutsche Bank a few years out at some set multiple of their revenues. Deutsche Bank was funding Frank's boutique, but didn't really own it. Since that wouldn't do, they quickly negotiated an exit for Frank and his team.

No problem, others on Wall Street wanted a technology presence. In June of 1998, CS First Boston, a perennial second-tier player on the Street, snapped up Frank and Associates, and gave them the same deal, minus the put. He was in control of all hiring and firing, of bankers, analysts, and even syndicate managers. I've seen the organizational charts—everyone reported to Frankie, who finally created the "boutique within a bulge bracket" firm he always wanted.

In a five-year deal, his group apparently got to keep 40% of revenues generated over $240 million (or some number close to this). Once again, Frank had a huge incentive to grow revenues at all cost. And he did. No Morgan Stanley-esque filter needed. If there was a deal to be done, they did it.

They also had a fund on the side to invest in private deals. There was a digital camera operating system company we were looking at, and Frankie was going to throw some cash in the deal as well. He told Fred and me to come in and talk to him about it. Their offices on El Camino in Menlo Park were two minutes from my house.

When you walked in, Frankie's office was the first one by the door. He could see everyone who was coming and going. Frank's clothing budget hadn't kept up with his compensation. As we walked to a conference room, I noticed a big rip in the back of his pin-stripe suit pants, and his wallet falling out.

"Hey Frankie, your wallet is so fat that it's ripped your pants."

I got a nasty glare, but I was right. As the boom took off, Frankie's group became an important source of revenue for CS First Boston. According to comments by CSFB's chief financial officer, from July 1999 to June 2000, the Tech Group was 12–15% of CSFB's $12 billion in revenue, or $1.5 to $1.8 billion. Even after subtracting the $240 million hurdle, 40% of that ain't too shabby.

Of course, lost in all this was the other 60% of all those tech group fees. Top management at CS First Boston had their hands in Frank's cookie jar as well, by setting up the structure to take their fair share, in their own bonus pool.

• • •

At Velocity, we stuck with technology companies, those selling software, chips or network equipment. To us, Inktomi was more interesting for their software than their search engine. We passed on the digital camera operating system company, but we kept on top of everything that was going on. We went to see every IPO road show we could stomach, both by Frankie's group and by everyone else. It didn't take long to realize that a daily lunch of rubber chicken was no way to go through life, especially when the companies were turkeys.

Even more annoying was that despite running a technology-only fund, we didn't get very many IPO shares allocated to us. Sometimes we got a meager 100 shares. Someone else was clearly getting shares in these deals. It pissed me off enough that we soon stopped going to CSFB's road shows altogether.

Many of these new companies going public were selling stuff on the Internet and worse than turkeys. Pets.com, Drugstore.com, Priceline.com, and Buy.com were all low-margin companies, but their stocks were valued at many times their revenues. As the ducks quacked, these stocks went up anyway.

CHAPTER 12

Price Targets as a Marketing Tool

Working on Wall Street from the West Coast is an ordeal, especially for those people like me, who like to sleep. Since I was in charge of trading, and the market opened at 6:30 a.m. California time, I had to be up and at 'em before my breakfast.

In December 1998, I flipped on CNBC as I do every morning for a dose of Squawk Box, to get a pulse of the market. You didn't need a call from salesmen anymore—all the market impact calls were delivered over CNBC before the market opened. Analysts figured out that CNBC was a better medium to reach clients than a sales force, and it didn't yell at you like a salesman would. I was skeptical of analysts using CNBC, because it was the old "I'm not paying for research that I can buy for 75 cents in the morning paper," except that it was on television and free.

That morning, as I dragged a razor across my cheek, the buzz was about Amazon.com. Henry Blodget, an analyst at third-tier CIBC Oppenheimer was pounding the table on the stock, raising his price target to $400. I started laughing and had to stop shaving so I wouldn't bleed to death. This was the single most stupid call I had ever heard. I hated price targets, because they were bogus. You used them when you have run out of anything else to talk about, but wanted the attention. Now here was Henry Blodget making a name for himself. I had heard his name before, but only in passing. He was a former journalist. Good for him, he has learned the game quickly, I thought. Amazon's stock popped 46 points, almost 20 percent that day. Now that's marketing.

I had just started dealing with Jim Cramer, another CNBC regular. I knew him slightly from my Morgan Stanley days. His web site called TheStreet.com, looked useful, so I contacted him to see if he'd be interested in a diatribe from Silicon Valley every once in a while. He jumped into action, putting me in touch with Dave Kansas, an ex-*Wall Street Journal* writer who was now editor of the site. I told Dave I would be willing to write a column every once in a while.

"Don't you write a column for *Forbes*?"

"I do, in sort of a pitching rotation, once every four issues."

"What do they pay you?"

"They pay me a dollar a word. I send them columns of 20,000 words, but they always seem to cut them back to 750 words."

"Well, we can't pay you that much."

"I'll do it for options."

"What do you mean?"

"You're a private company. Don't pay me a penny. Do it the Silicon Valley way, pay me 750 options per column. I'm willing to take the risk that they will be worth something someday. But hurry up, you can only give out cheap options a year or more before your IPO. The accountants and the IRS will hassle you about cheap stock when your IPO is near."

"Whatever."

This was the summer of 1998.

In December 1998, a few weeks after Henry Blodget's price target splash, TheStreet.com decided to do a webcast of an investment panel. In addition to Jim Cramer and Dave Kansas, columnist Herb Greenberg would represent TheStreet.com. Outside panelists were Henry Blodget from CIBC, Ryan Jacob from the booming Kinetics Internet fund, a guy from Munder whose NetNet fund was also on fire, and me with a view from Silicon Valley.

I flew to New York to take part, breaking an important rule of mine to never travel to New York if snow could possibly be in the forecast. It snowed.

The panel was a lot of fun. Henry pounded the table on Amazon and Yahoo and other "quality" Internet names. I told investors to stick with infrastructure companies, those that sold network equipment, chips, and software to these dotcom companies. Ryan Jacob, who looked about twelve and acted even younger, had some twisted views on spotting advertising trends and page views per share. I decided I wasn't going to invest in his Internet fund.

I liked Henry. He could talk a mean game. I had been in that game of pitching the same stuff he was pitching, so could easily see the thorns from the roses. That's OK. He believed in what he was saying, and conviction is important on Wall Street. You rise to the top if you can help portfolio managers buy shares by providing them with your strong conviction. Henry would do well.

In fact, as the year turned to 1999, he was hired away from sleepy CIBC to Merrill Lynch, which was losing the Internet investment banking battle to Morgan Stanley, Goldman Sachs and Frank Quattrone at CS First Boston. By hiring Henry, they were making a statement that they were players. Merrill Lynch needed all the help it could get.

In the midst of growing into technology players, there was even an effort to buy the boutique Hambrecht and Quist. It got shot down by internal politics and some talk that H&Q was doling out hot IPO shares to investors who later did banking business with them—so-called spinning. It smelled like Merrill Lynch wanted to emulate Frank Quattrone, and have a "boutique within a bulge bracket" firm. Now Merrill had Henry Blodget to peddle to dotcom companies.

In 1998 and 1999, there certainly were enough deals to go around. The trick was to have a visible enough analyst who could impress upon companies that they would support them in the market after the IPO. It was all about marketing. One hundred phone calls a month, the "Ohio Death March," "Sherman's March to the Sea," Metroliner to D.C.? Nah. That was too much trouble.

Merrill Lynch hired a public relations firm instead. They got the best in the business, Pam Alexander at Alexander Ogilvy. I have known Pam since the 1980s and would run into Henry at various events and dinners Pam hosted. I offered him some advice, on how to make *I.I.* (as if it mattered anymore), how to keep on top of stories, get stock picking right and everything follows, hold off the bankers, and even some thoughts on marketing to institutions versus the retail crowd at Merrill Lynch. While he seemed grateful for the suggestions, I'm not sure he was terribly pleased to be offered advice by anyone. He was on a roll.

Then it hit me between the eyes. The mold of a successful analyst was broken. Getting ahead by working feverishly with discriminating institutional investors until your reputation and status were proved—well, those days were over. The ducks didn't care about reputation—they simply wanted stocks that were going up, and analysts to pound the table to keep them going up.

• • •

Fred and I would laugh at the pieces of garbage purporting to be companies that were going public, and calculate how much of the $100 million that they raised would be spent on switches, routers, and software. We were doing well NOT investing in dotcoms, but staying down the food chain.

My biggest laugh was when I received an issue of Barron's at the end of 1998. Mary Meeker was on the cover, and labeled the "Queen of the Net." Fred remarked to me that most Queens eventually get beheaded.

In order to keep up with Frank at Deutsche Bank/CS First Boston and with the bottom-sucking Goldman Sachs, Morgan Stanley's filter was now Swiss cheese. You couldn't trust boutiques and now you couldn't trust the "boutiques within bulge bracket" firms.

Momos didn't care. They quacked and quacked. The more hot IPOs they bought, on the deal and for months after they went public, the better their performance. The better their performance, the bigger the ads they could run in the *Wall Street Journal* and the more money they would take in. They just needed more deals to feed into their insatiable machine.

Momos didn't care, but I started to get concerned. Fred and I had stopped going to road shows, even Morgan Stanley's. I forget whether it was Ask Jeeves or Akamai, both of which were Morgan Stanley deals in July 1999, but it wasn't funny any more. They were not "Morgan Stanley material," as Frank Quattrone used to put it, let alone companies that anyone in the public should own. Morgan was now in that same bar late at night, and had lowered their standards too.

Mary made the cover of *Fortune* magazine as the third most powerful woman in business. Now that's marketing. *New Yorker* magazine did a profile on Mary. One of the quotes caught my eye:

> *"My view of Amazon is that it's not just books, it's bits. If two years from now it has fifteen or twenty million customers, and it has their credit card numbers, and they are happy, then it can make money."*

I can hear her mind going through it, "I'm happy. Customers will be happy. The stock is going up." Same old Mary. It was just what the market and the ducks needed to hear to justify higher stock prices. Eternal happiness.

The ducks spread beyond stock funds. The bond guys wanted a piece of the action, too. Morgan Stanley and Mary Meeker got back in the good graces of Amazon.com. They suggested to Amazon that their stock would be going higher—Mary Meeker had a Buy rating to prove it—so don't sell more stock to the public, sell a convertible bond. The pitch was simple: your stock is around $150, so sell $750 million of convertible debt to investors that converts into stock at $225 per share. The only way you will ever have to pay back the debt is in the highly unlikely scenario that your stock trades below $225 in a few years. Oops. (Don't compare these numbers with the Henry Blodget $400 price target because the stock has split a number of times.)

Morgan Stanley loved these converts because they could make money trading them. Spreads on over-the-counter stocks had gone away. If Amazon had just sold more shares to the public (as they should have), Morgan would have not made much money on their trading desk. But by trading more than $1 billion of convertible debt (Mary had apparently told Amazon to increase the amount they raised) with fat spreads, Morgan Stanley would make tons of money on their trading desk. Companies fall for this convertible nonsense all the time. Beware of whose agenda is being fulfilled.

• • •

In 1999, research from the Street slowed to a trickle. Not that we used it anyway, but because it was the pulse, we had to see what analysts were saying. Not much, it turned out. They were all on the road shows with management pitching new deals. Instead of research, our office was flooded with prospectuses for new deals. We would pile these red herrings on a shelf by our only window until they finally blocked the sunlight. Eventually, we just took them from the mailman and immediately tossed them out.

Not five years earlier, Morgan Stanley had four technology

analysts with Frank Quattrone and a small crew as technology investment bankers. Now they had over fifty analysts covering every technology industry segment, including "periph-ree-als," and over one hundred investment bankers in their Sand Hill Road office in Menlo Park, California, alone.

I struggled with what made this whole system tick. I had friends who had growth funds with $10 billion, $20 billion, even $40 billion in assets. They did absolutely no fundamental research. No Hank Hermann-like Piranha tactics to figure out what analysts were saying and how the market would react to the next piece of news. They just sat in conference rooms and had IPOs pitched to them. And they bought them all, and rode 'em hard.

• • •

Our fund was an investor in a few of these deals. I must say, there is nothing more gratifying than a hot IPO. Nothing beats the call from a trader saying, "We're looking at an indication of $40–45." How cool is that. We paid $4 in a venture round, the IPO price is $16 and now it's a ten-bagger. Let the champagne corks fly. An instant fortune, right?

Well, not really. Wall Street insists on a six-month "lockup" of shares. Only those who got shares on the first day could sell them. Entrepreneurs, management, venture capitalists, and employees all had to wait around for 180 long days. Fair or unfair, this silly lockup was one of the roots of the IPO boom.

IPOs are best analyzed by the old Wall Street adage that describes the process: "Everybody gets paid." Underwriters collect 7% fees, the last fixed fee area on Wall Street. Before things changed and all deals were hot deals, the way IPOs were sold was to price the stock at a 10–15% discount of a similar company's current value. This was so buyers of IPO shares would get 10–15% first-day gains. That's it. That covered the risk of buying shares in a stock that had never been public. Some would go higher, some lower, but buyers were enticed by these 10–15% gains.

The company going public, despite paying 7% to bankers and another 10–15% to buyers, got cold hard cash in exchange for brightly colored stock certificates, at a much higher value than they could get in the private market. Venture capitalists are not often generous souls.

Entrepreneurs and the existing venture capitalist investors are supposed to get an orderly market for their shares 180 days down the road so they can leverage their sweat and cash. America gains by encouraging a higher level of risk-taking and innovation that is lacking in the rest of the world.

But somewhere in this mix, the system broke. Shares are popping 50–500% on the first day, not 10–15%. Everyone wanted in. Barbra "Legs like Butta" Streisand was calling around trying to get shares.

Fund managers promised to buy more shares in the open market on the first day of an IPO, to *ladder* the deal, causing or perhaps just perpetuating the first-day pop. In the rare instance that our fund had to buy more stock in order to get a decent IPO share allocation, we did promise to buy stock after the IPO started trading. But we set a price limit. If the IPO was priced at $18, we said we would buy more shares up to $24 a share. When the first trade was $42, we were off the hook.

Because almost every deal was practically guaranteed to go up, IPO shares were *spun* to CEOs, board members and venture capitalists who could sell immediately, make instant money and would pay back Wall Street with deals down the road. I used to get tons of calls from friends in the business asking about some recent IPO. "What do you think of the company? Is the stock a keeper?"

"Not much, blow it out," was my usual answer. I'd then ask, "Why do you want to know. I didn't get any shares, did you?"

"A few. Thanks. Good-bye."

While CS First Boston was not the only firm allocating hot IPOs, the Valley was buzzing with stories of "Friends of Frank" accounts. You were the ultimate insider if you had one.

Finally, and most disgusting of all, it appears that certain investment banks were collecting outsized commissions from sleazy clients, *kickbacks* if you will, to share the instant gain. Ah, so that's why my fund wasn't getting shares, "Friends of" and these fast-money hedge funds were paying under the table for them.

IPO shares became like Ivan Boesky's cash in briefcases.

. . .

So, how did it get out of hand? Well, the investment bankers are supposed to be underwriters. They buy shares from a company at night and sell them to the public the next morning. For the "risk" of this transaction, Wall Street collects their 7%. What a joke—there is no risk to IPOs. If they can't sell the deal, bankers call it off.

A carefully orchestrated two-to-three-week road show whets the appetite of investors. Management feeds prepackaged dog food presentations to investors who gobble rubber chicken lunches. Five (out of fifty) million shares are offered, the other forty-five million shares are "locked up" for 180 days. That's cheating. Supply and demand is out of whack for 180 days. Momos are quacking so loudly, demand is huge, but with only those five million shares to satiate them, investors fight over them like real ducks fighting over a piece of bread. If all fifty million shares are trading, they would never have traded up so high, and momos would be out of work. The whole damn thing is rigged by these lockups.

But lockups are what Wall Street bankers insist on to create an orderly market for IPO shares when none existed the night before. Tough. Wall Street doesn't need to rig the market like this. You can blame the bubble on greedy bankers, venture capitals, fund investors, or even Alan Greenspan for over-pumping the money supply. Fine. But lockups are the structural problem that kept the right amount of food from satiating the ducks and instead pumped a lot of hot air into the bubble, and no one wants to admit it.

• • •

In 1999, Fred and I had the fourth best hedge fund in the U.S. Better lucky than smart, I suppose, because timing is everything. Our performance numbers were not even close to being believable, and more surreal than real, so we started returning money to our investors. Was it because we thought valuations were out of whack? Analysts conflicted? The IPO market insane? Lockups rigging the market?

I wish I were smart enough to point to one of these and say we called the top. Not a chance. We drank the Kool-Aid, and believed the bullshit, too. After struggling for years to raise capital to get to $100 million in assets, we found ourselves in the summer of 1999 with $1 billion in capital. Hey, we must be "fookin' geniuses," as someone on Monty Python might say. It was no longer hard to raise money. Instead, people found us.

In July 1999, Fred got a phone call from someone on Wall Street who had been helpful to us. "Can you guys please meet a client of mine for breakfast tomorrow morning? He has heard about you guys and is interested. I know you probably have too much money now, but could you just do me this favor?"

So we trudged over to Il Fornaio, one of three Silicon Valley money power breakfast locations (Buck's and Hobee's are the others). We showed up early and waited out front, waving to John Sculley who we saw just about every time we were there. A few minutes later, the biggest, monster stretch limo pulled up, and eight well-dressed men climbed out. Three of them could have played linebacker for the Oakland Raiders. They turned out to be bodyguards. We took a table at the back. (Hey, wait a second—haven't we heard this story before?)

In fluent English, the leader of this group explained that they were from Bahrain. They had seen our performance numbers published somewhere and wanted to meet with us. He was very gracious. Fred and I told our story, small cap public, later stage private. We hadn't got far with it when he interrupted us and said, "I like what I am hearing and am very interested in your fund. I would like to propose that we wire $500 million to you tomorrow morning."

I think I passed out for a second, but Fred had the sense to speak up and say, "We are flattered but we do have $1 billion and we invest in small companies. There are only so many of them that we can find that generate the returns we expect. It was nice to meet you and thank you for coming." I almost kicked Fred under the table. All that money and Fred was ushering them out? The fees from that amount of money was nothing to sneeze at, and my kids needed new sneakers and Pokemon cards.

Back at the office, I said I thought we should take their money. What's the big deal? But Fred explained that if we take it, we would work for them, not for ourselves, and that we really did have too much capital, that we still might be fookin' geniuses, but we didn't need their money. Fred was right.

A week later, I received a call from another Wall Street person we owed a favor to, requesting that we meet with a client of his the next morning for breakfast. I explained that we probably weren't going to take anyone's capital, but he was persistent, so it was off to Il Fornaio again for breakfast.

This time, after a wave to John Sculley, a normal-sized stretch limo pulled up, and four guys piled out. We took the same table at the back.

"We are from Saudi Arabia and we have heard about you guys and are interested in your fund." Without waiting for us to describe what we do, he added, "We are prepared to wire $500 million into your fund tomorrow morning."

Is $500 million some sort of Middle Eastern unit of money? A shiver went down my spine. But I didn't pass out this time, and immediately jumped in and said, "We are flattered by your generous offer of capital for our fund. You know, we invest in small companies and there are only so many of them." And on I went.

This time on the way back to the office, I started babbling and told Fred that he was right about not taking last week's money and we were right in not taking this week's money. But, someone was going to take it. That $1 billion was going to find its way into venture capital funds and hedge funds and was going to muck up the market one way or another. Not only should we not take any more money, but we should also start sending back the money we already had.

Fred started laughing. "That's exactly what I was going to tell you."

Of course, we should have sold everything that day. We didn't. After all, we were fookin' geniuses and could make great investments better than anyone. But just in case we couldn't, every month for the next two years, we called up investors and threw them out of the fund, or told them they had to take back a third or half of the capital in their account. There is yet another old Wall Street adage that says, "There is no bell that rings at the top (or bottom)." Market tops and bottoms are always figured out later on. We had the next best thing happen to us. The limos and bodyguards and money thrown at us were a tiny bell that saved our butts.

CHAPTER 13

Synthetic Goldman Sachs

In the midst of this raging bull market, Wall Street was changing. You couldn't tell, because banking fees were so huge and trading was so rapid that profits at Wall Street firms kept going up. Of course, so did compensation. Our fund was doing well, perhaps because we had cut off Wall Street research. I knew how that sausage was made and couldn't stomach paying for it let alone eating it. My partner and I invented our own research call, driving around Silicon Valley, looking under rocks for small companies that might someday become big players and get research coverage from those "boutiques within bulge bracket firms." I was volunteered as the head of trading operations (I was also CFO, janitor and water boy.)

Trading with the Street is a perilous activity. It is easy to get ripped off. There were so many new companies going public that only the underwriters of a company would make a market in each stock. "Make a market" is over-the-counter trading lingo that means provide a Bid and an Ask. If you wanted to buy 10,000 shares of a company, you had to call up the investment banking firm, and they would "work" the trade. This meant they made a few calls to see if anyone had any for sale. If so, they would try to buy it for you, marking it up an eight, a quarter, whatever they could get away with. I opened an account with firms based on my need to trade stocks that they had taken public.

I stopped dealing with salesmen and saleswomen almost immediately. They would call with their spin on what their analysts were saying, but I had already received an email that morning with their morning call notes, or had seen it on CNBC. "Save your breath," I would tell them.

Since we were buying pieces of small companies whose shares didn't trade very often, buying a big enough position of, say, 100,000 or 250,000 shares would literally take months.

I insisted on the phone number of the trader. I wanted direct access. I figured out once that I was spending 1–2% in commissions every time I got in and out of a stock. Many times, shares would be for sale, at say, $10¼ and I was willing to buy them at $10¼, but traders needed to buy them at $10 and then mark them up a quarter to sell them to me. No markup, no trade, even though the market was $10¼. That would piss me off to no end. The reply was always the same, "Nothing I can do for you, gotta make a living."

I was working an order with a sales trader at Robertson when she casually mentioned, "I'm not finding much for sale anywhere, but I have been buying it off the box for you."

"Great, thanks." I had no clue what buying it off a box meant. More trader lingo I didn't know about?

A few days later, risking perpetual embarrassment at being a trading rookie, I asked, "What box are you buying it off?"

"Oh, the Instinet box," she offered.

"Oh, yeah, sure, of course."

I called up Instinet the next morning. "How do I get me one of them boxes of yours?" I learned that Instinet is a matching service. Buyers and sellers put in bids and asks and anonymously buy and sell shares to each other for three cents a share, maybe two cents or a penny a share if you did enough business. Robertson was buying shares for me for two cents, and then marking them up 25 cents to sell them to me. What a great racket.

Within a few weeks, Instinet had installed a machine and a private communications line into their system and voilà, I was buying shares myself, for pennies. And I was in control of pricing. If I wanted to move my bid up or sit tight and wait for someone to bring their ask down, it was all up to me sitting in front of a screen instead of making 20–30 calls a day to the same trader. So we had successfully cut off analysts and salesmen, and now, with Instinet, traders as well. What did we need Wall Street for?

• • •

Remember that scene of the market crash in 1987, and traders not answering their phones? It started a bunch of dominoes falling. The Securities and Exchange Commission insisted on the implementation of a system called SOES, or Small Order Execution System. Trades under 1000 shares would be executed automatically at the current market price.

Two smart programmers, Jeff Citron and Josh Levine wrote an MS/DOS program on their PC that could game the SOES system, electronically sending in rapid-fire trades to pick off over-the-counter traders. These guys were not so affectionately known as SOES bandits, and roadblocks, including limits like only one trade every five minutes per trader and only so many per day, were put in to slow them down.

This led to the creation of day traders—whole rooms filled with SOES bandits to get around the per trader rules. When a trader had a big block of shares to buy or sell, it was impossible not to run into these bandits and lose your shirts on these trades. So traders would often signal to others on the Street to get the hell out of their way, they had a big trade to cross, and other traders would pull their bids and asks to make room for the other guy to maneuver around the SOES bandits.

In 1996, the day trader loophole led to a massive domino dropping in the middle of every trading floor, a class action lawsuit called the "NASDAQ Market-Makers Antitrust Litigation" led by William Lerach. It alleged collusion amongst Wall Street traders. In 1998, Lerach proved his case. Of course he did, since they had to collude to get around day traders. Wall Street settled for $1 billion. The SEC also put in a regulation, Rule 11Ac1-4, the Limit Order Display Rule, authorizing Electronic Communications Networks or ECNs. These were computer systems that could match trades without human intervention.

It was not just the loss of market share to these anonymous matching systems that would alter Wall Street. The Lerach suit and its $1 billion settlement killed liquidity, which is what makes markets work efficiently. If you want to sell shares, but there are no buyers around just then, someone like Goldman Sachs might step up and buy your shares for their own account, and then sell them over the next day or two. Goldman Sachs would use its own capital to facilitate your trade. Wall Street legend tells of a sign on the football-field-sized trading floor at Goldman Sachs that read, "If you can't find the other side of the trade, you are it." That is liquidity. It kept stocks from bouncing wildly. Their payoff was the spread between what they pay for it, the Bid, and what they sell it for, the Ask.

Lerach sued because the spread was too large. Paying $1 billion got Wall Street's attention and sure enough, liquidity disappeared. If Wall Street couldn't get paid enough to do your trade, then they were not going to put their own capital at risk. I saw that change in 1998, when brokerage firms no longer bought my shares outright, but would instead "work" the order for days, even on bigger names that traded a lot. It was as if Goldman Sachs had put up a new sign reading, "If you are the other side of the trade, you are fired."

Without liquidity for over-the-counter trades, you could breathe on a stock and make it move. A big buy order was almost guaranteed to make a stock go up two, three, even five points. Likewise, a decent sell order could halve a stock. Perhaps that's why Amazon jumped 46 points on the famous Henry Blodget $400 price target call.

Instinet, now owned by Reuters, had been around for decades, but quickly turned itself into an ECN trading system. The same guys who started the whole day-trading phenomenon, Citron and Levine, tweaked their code and created their own matching system to compete with Instinet. They called their ECN the Island. ECN trading systems would do half of over-the-counter trading volume by 2001. Those traders should have answered their phones in October 1987.

I was hooked on doing my own trading, so I called up the

Island and opened an account with them too. There was no need for a box, because it worked off a secure web browser. I was in business. My share of trades with Wall Street firms went from 100% to 25% within a few months. You still had to pay the Street, for research (yuck) and to get invited to their conferences. Also large trades could still only be done with human intervention. But electronic trading was here to stay.

It also got me thinking. Wall Street was turning horizontal, and breaking into thin layers. There were already a few banking-only boutiques. Now there were trading-only operations. How about research-only? Could you really create a virtual Morgan Stanley or a synthetic Goldman Sachs as I had postulated about IBM years before?

• • •

The answer was a resounding maybe. It sounded pretty cool, so I thought I'd sniff around anyway.

I never did get cheap options from TheStreet.com, the website run by Jim Cramer. There was no one there who had any Silicon Valley savvy when it came to pricing options. Instead, they waited until the week before the company was ready to go public and then distributed options to everyone to whom they had promised them. I got a bunch, with the right to buy shares in a year or so at $9 per share, which was about where the IPO was to be priced. They weren't convertible for a year or more, but I figured that when I got $9 options instead of 10 cents per share, they might end up worthless.

However, I did end up as a direct investor in TheStreet.com. They had done a few rounds of financing before their IPO and I was happy to invest. It was fun signing papers because my name was sandwiched between Dave Kansas and Henry Kravis of KKR. A motley crew.

I was intrigued with the site. In 1999, it was clear to everyone on Wall Street that research was a façade to support deals. It was the open secret on the Street. Everyone knew it, but deals were being done, fees were being generated, ducks were quacking, institutional clients weren't complaining, retail investors were making money. Why rock the boat?

But someone could do research right.

• • •

TheStreet.com was getting ready to go public. Goldman Sachs was the lead investment banker, I suppose because Cramer had worked there for many years.

I learned a trick from Bob Harris. Whenever a company goes public, they almost always try to "lock up" all the existing shares of a company, so that when they sell new shares, all those existing shares don't come crashing into the market and spoil the stock market debut of the company. A lockup agreement is distributed to management, venture capitalists, and any other investors in the private company. By signing it, you agree that you won't sell your shares for 180 days after the first day of trading of the initial pubic offering.

I hate lockups. Let me take that back. I loved lockups when I was in the investment banking business, as they helped create hot IPOs that trade up spectacularly and allow traders to control trading for six months. As an investor, I hate them. The stock pops in the first days and weeks and I can't sell!

Bob Harris taught me that when the lockup agreement arrives, you just put it aside on your desk and wait. Eventually, someone from the underwriter, or from the investment banking firm taking the company public, will call and ask for the signed lockup. Then you ask, "What are you going to do for me?"

Goldman Sachs sent the lockup agreement for TheStreet.com, and I dutifully placed it on the side of my desk. Sure enough, the days before the deal was to be priced, an associate from Goldman Sachs called asking for me to sign it and return it.

"We can't proceed with the deal without your lockup."

Now, I didn't own that many shares, so that statement was highly doubtful. I said I wasn't going to sign it. There was nothing in it for me. It moved up the chain, a banker called, and then a syndicate person. They got the same answer from me. I was waiting for the head of syndicate who allocated the shares in the deal to call so I could ask for a 50,000-share allocation in exchange for the lockup. It never came. I was free and clear for the first 180 days.

The stock popped to $60 the first day, valuing the company at $1 billion. I should have sold there and then, but felt a sense of loyalty to the company. Stupid me. That was the highest the stock would ever see. The ducks were getting tired.

I was still intrigued with the site as a forum for independent research, so I placed a call to Jim Cramer to set up a meeting. I told him that I had an idea for how his site could do just that. Wall Street was going horizontal. Someday, there might be separate banking firms and research firms and trading firms, and this was an opportunity for his company. And hey, I was an investor. I wanted them to get it right.

I showed up in his office on Wall Street and met Jim and his partner Jeff Berkowitz. There were a few small offices and a 20-foot by 20-foot trading room, with five people running around and yelling into phones. I chuckled since my firm had my partner and me. That's it.

"See those people in there?" Jim asked.

"Yeah."

"They're holograms. They don't do a damn thing. We can't leave them alone. Luckily, there's not much going on, but I only have a few minutes and then I've gotta get back to the desk."

Berkowitz just shook his head saying, "Goddamn holograms."

Unbeknownst to me, Jim had invited two of his board members, Fred Wilson and Jerry Colonna, from Flatiron Partners. I had known Fred from his previous venture firm, and liked him. He was very personable, smart, and seemed to think things out. I had only heard bad things about Jerry—that he was a journalist turned venture capitalist and was making stuff up as he went along. These stories were right. Not that it was a bad thing for a while. Flatiron had funded GeoCities, that skanky Goldman IPO, which was sold for tons to Yahoo.

I told Cramer that Street research was conflicted. He nodded agreement. Mary Meeker was an invention of the system to sell deals. Cramer got up from his chair and started pacing the room. Investors needed unconflicted research, and would pay for it, in commissions or soft dollars, but probably not advertising. He started flinging pencils over my head. Wall Street was going horizontal, and would need independent research. Amen. TheStreet.com didn't need to hire a Mary Meeker—they needed to invent one themselves. Cramer started grunting loudly. It wouldn't even be all that hard. Jerry disagreed. Fred thought out both sides of the argument.

Cramer's pacing became faster, and he started talking out loud to no one in particular, thinking it out, talking it through, what the page would look like, two columns, morning call on the right, who they might get, what were the negatives. "Yes, yes. This is exactly what we have to do."

Jerry thought otherwise and probably killed it at the board level. The way TheStreet.com got around the potential conflict of Cramer running a hedge fund was that he wasn't allowed to be a part of management. He was just a contributor—and a board member with one vote. Instead, the company expanded in London, spending $15 million setting up the office and site and sales force to sell ads. They could have hired 30 Mary Meeker clones, one of whom would surely have hit.

I sold my shares. No lockup.

• • •

Others came in to fill the vacuum. Independent research has always been around, but in 1999, it blossomed in many places. One independent voice belonged to George Gilder, author, futurist and now newsletter writer working out of the Berkshires in Massachusetts. George used to write about economics and politics, but got bitten by technology and wrote his best-selling book, *Microcosm,* about the semiconductor industry. I read it with interest. Hey, it was my industry—I had to know everything about it.

One day, back in my time at Morgan Stanley, economist Steve Roach had called me and said he had this guy, George Gilder, on the phone. He had some questions about Motorola and CDMA, a new type of cell phone protocol. Sure I'll talk to him. I told George. "As you know, I think that CDMA is a few years out and Motorola is not done milking the analog cell phone business. Plus, CDMA comes out of this little company in San Diego named Qualcomm and I doubt Motorola would use it, since they didn't invent it." OK, I didn't say, "As you know."

"Too bad, CDMA is going to be big. I just figured it might be next year," George replied.

George Gilder was right, but five years too early. This would be a recurring theme. We exchanged phone numbers and emails and kept in touch over the years. He even invested in our Velocity fund.

In 1999, I was reading the morning *Wall Street Journal,* in need of my fix of the pulse. The "Heard on the Street" column was about how George Gilder had "recommended" a number of stocks the day before. Each one of them had jumped 10–25%. He was an ax. While intoxicating, this was not good. You don't want to be an ax if you don't even know what an ax is. George was treading into unknown territory. I picked up the phone and called him.

"George, are you a registered representative?"

"A what?"

"Did you pass your Series 7 test?"

"I'm not even sure what that is."

"Did you know that you have to pass this test before you can recommend stocks to anyone?"

"But I don't recommend stocks."

"It doesn't matter if you do or you don't. The *Wall Street Journal* is the paper of record. If they say you recommend stocks, then you recommend stocks."

"But . . ."

"Look, what I would do is immediately put a disclaimer in your newsletter. Something like, 'These are not stock recommendations.'"

"We can do that."

"You are an industry analyst, not a stock analyst. Make sure your readers know this. Talk about fundamentals, not stocks."

"I do."

"Look, I recommended stocks for years. You don't want to do it. It sucks. When they go up, people love you. When they go down, they blame you. It ain't fun."

"I understand. Can I bounce a disclaimer off you when I come up with one?"

"Just get any Morgan Stanley or Merrill Lynch research report. They have $1 million-a-year lawyers to write their disclaimers. Steal one of theirs."

• • •

Hey, I ran into an old name. In the summer of 1999, I was asked to go look at a supposedly hot company, Be, Inc., run by none other than Jean-Louis Gassee, the guy I met at the Apple analyst meeting back in 1985. He was friendly enough this time. He needed to raise money, which makes everyone nice. When I asked a few tough questions about the details of his system, he shrugged his shoulders and said "I dieu nut kneeaauuww." And instantly reminded of that 1985 meeting, I knew he hadn't changed much. What goes around comes around. I didn't invest. Be eventually became "Not to Be." It wouldn't be the last time some ancient memory kept me out of trouble.

A few weeks later, an old college friend, Jarett Wait, who

worked at Lehman Brothers, asked if I wanted to go meet a friend of his that runs Buy.com. Still private, they were billed as the next Amazon.com, selling books and computers and other stuff at prices less than anywhere else.

"Where are they located?" I asked.

"In beautiful Aliso Viejo," he answered. "In that no man's land between LA and San Diego."

So I met Jarett at John Wayne Airport, and we drove down to Buy.com, conveniently located across the street from a United Parcel Service building. Scott Blum, Buy.com's CEO, met us in the lobby and took us on a tour of what basically was a huge warehouse with some computer servers in the back room that set prices just below Amazon.com and rung up orders. I think they wheeled packages across the street to UPS a few times a day. We filed into a conference room.

"We did about $10 million in sales our first year, and $100 million last year. We ought to hit $1 billion this year and $10 billion next year," Scott Blum explained.

"Wow" was about all that came out of my mouth.

"We've met with everyone on the Street. Blodget and the Merrill folks are interested in taking us public and so is Meeker and Morgan Stanley. Goldman and CSFB are coming down next week. They all tell me the same thing, that the company will be valued at the IPO at around three or four times our current sales, and maybe trade up to ten times sales in a good market. John Sculley is on our board and is helping us pick bankers."

"What kind of margins do you have?" I asked.

"Not much, but if you make 1% on $10 billion in sales, that's huge profits," Blum explained.

I got up to get some coffee, and poured half and half into my mug. The carton said Foremost or Knudsen. I'm not sure since it started getting fuzzy. I had a major flashback to my first analyst meeting back in 1985, at the milk company that had billions in sales. If I remember correctly, it was only valued at $100 million because they didn't make much money, their margins were barely 3%. Buy.com wasn't much different than the milk company. They were in the same distribution business. They weren't worth ten times sales, more like 1/10th of sales.

"Good luck, gotta catch a plane out of here," I said.

CHAPTER 14

The Ax Syndrome

In 1999, I met up with Jack Grubman at the Salomon Brothers Telecommunications Conference at the Waldorf Astoria. I spent three days watching him in action, and we spent a bit of time standing around in the halls laughing about old times and all the funny stuff we did when we were both new to the game.

Jack was a Wall Street superstar. While we were talking, every investor at the conference came up and paid homage to him.

"Great conference."

"I want your thoughts on that company."

"What's the next big trend, Jack?"

"I just want to make sure we get a big piece of the next deal."

Hmmmmm. I heard that last one a lot. Clients were marketing to Jack. Salomon, because of Jack's relationships with management, was doing a lot of deals. Allocations on IPOs are always skimpy, but here were clients positioning themselves to get bigger allocations. I doubt that Jack had any influence on syndicate allocations to carve out bigger pieces of IPO shares for institutions, but who knows. Jack, of course, was eating it all up. I got a few "can you believe this shit" looks from Jack.

I just blurted out, "Man, you have a lot of people believing your bullshit."

"Part of the job, my friend, part of the job."

I tried to explain a little bit of what we were doing with our fund. Jack quickly turned back to the self-marketing, going through a list of private investments that cable mogul John Malone allowed him to put money into. It turned out that I had looked at a few of them, and passed. They were a doggy list of companies. Malone was not doing Jack any favors by letting him invest in these deals. I kept this opinion to myself—why offend an old friend?

One night of the conference, Jack hosted a dinner for Alex Mandl, an old friend of his from AT&T. I remembered meeting Alex when Jack brought him into Paine Webber some years earlier. Mandl had just signed on as CEO of Teligent, a company that owned a bunch of radio spectrum, basically frequencies they would use to bypass phone companies. They would buy two $5000 radios, put one in their facilities and another at their customer's, and wirelessly create a data connection. These T1 lines were going for $1000 a month from phone companies, so Teligent could charge just under that and get paid back for their radios in less than a year.

The dinner, at the Gramercy Tavern on 20th Street in Manhattan, was very exclusive. A dozen or so of the top institutional investors were invited, and Jack asked me if I wanted to tag along. In a very private room, over lobster canapés and New York strip steaks, Mandl gave his pitch. It was modestly interesting, if you thought T1 lines were going to stay at $1000 a month. For Jack, it was a coup. He was introducing Mandl to future buyers of his IPO. As the dinner ended, Jack stood at the door with car service vouchers, handing them out like candy. He had prearranged for a dozen Lincoln Town Cars to drive dinner guests home, whether in the city or out to Connecticut. What a nice service of Salomon Brothers.

I stuck around at the bar for another drink with Jack and Mandl. It wasn't hard to figure out that Jack was raising a private round for Teligent, and that Mandl's job was to precondition the market for his IPO. By telling private investors that they already had a dozen of the top institutions ready to buy an IPO, it was not hard to raise private money. Jack seemed to be orchestrating it all. It's not that there was anything wrong with that. Raising money is a pain in the neck, and requires the finesse of playing poker when you are dealt a Jack-high hand.

During the rest of the conference, my mind was filled with self-doubt. Did I do the right thing by hanging up my analyst cleats? That could be me running a conference and lining up money for companies. Watching Jack and Mary Meeker be the toast of the town was awkward. Of course, I could have easily screwed up and been a bit player on Wall Street, too. Screw it, I realized, I was having more fun in a dumpy office in Palo Alto. Being an analyst was just no longer worth the aggravation. It is great to be visible, but only if you were right. I remembered that lesson well. I quickly got over it.

Six months later, our Salomon Brothers salesman called and said they were initiating coverage of Teligent, a company they had taken public a month before, with a Buy rating. Now *that* made me laugh. T1 prices were already caving in as fiber optics were being laid by other Salomon Brothers deals like Worldcom and Qwest. Teligent was toast. No matter, Salomon made a killing on fees from raising the private round, from the IPO, and probably from a bunch of debt deals to come. Eventually, John Malone's Liberty Media, then controlled by AT&T, would buy Teligent and they would sink together. Telecom was a regulated business, and everyone, I mean everyone, thought prices would never go down. That idea was deadlier than the bubonic plague. Regulated prices didn't go down, but so much capital was thrown at the telecommunications industry that the new players all competed with each other, and prices eventually collapsed. Economics 101 stuff. Analysts and bankers were too busy collecting fees to figure this out.

• • •

Jack Grubman was a rare bird on Wall Street. From his Paine Webber days, he had been trained to make money for clients, by picking stocks that went up and avoiding those that went down. He worked hard at it and he was the best at it.

But he also was a friend of management. Like that old pay phone competitor that opened his eyes to AT&T competition, Jack figured out you could make a killing betting against AT&T. Worldcom CEO Bernie Ebbers was the ultimate anti-AT&T, and Salomon did deals galore funding Ebbers acquisition spree. Jack even raided AT&T, helping place his old AT&T buddies into CEO slots. Alex Mandl at Teligent. Joe Nacchio at Qwest. And a bunch of others.

They all used Salomon for investment banking because they all knew that Jack had the ability to sell deals better than anyone. Investors made money by him, and trusted him. He was an industry ax. One word from Jack, and billions could be raised. This was no overnight success story. Jack had cultivated this position for over a decade and it was paying off.

Salomon had to keep him. Goldman Sachs would have loved to hire Jack, and from what I understand, they tried a few times. So Salomon had to pay him more as he brought in more deals. Remember, Wall Street is a compensation scheme. Half of revenues and fees are paid out as compensation to employees. Jack was generating revenue, and it was increasingly easy to justify his piece of the action.

But the ax stuff, as I mentioned, is amazingly intoxicating. It is almost impossible not to get drunk on the glory of moving stocks and of raising billions. Jack became a kingmaker. Gary Winnick had started Global Crossing to sell undersea cable and take market share from AT&T and others. His first undersea cable, Atlantic Crossing 1, cost a few hundred million and was soon generating $100 million a month. Jack was just the guy to help raise billions to duplicate the effort, one more time in the Atlantic and then under the Pacific.

• • •

The dotcoms' meltdown started in March and April 2000. Most bubbles end when no new money can be found to keep blowing hot air. Whatever the revenue numbers or margin numbers or earnings forecasts companies reported, their stocks went down. The ducks were quacking in the opposite direction. Momos started dumping. Until the fall, it wasn't obvious that it was all over and many investors kept looking for signs that things might turn back up, or that there might be a bottom in these stocks.

Bob Metcalfe, the inventor of Ethernet networking and a networking industry legend, ran a conference called Vortex, and he asked me to be on a financial panel that he was moderating. In May 2000, I caught a flight to John Wayne Airport in Southern California and then cabbed it to the Ritz Carlton at Laguna Niguel. I had been there a million times for various investment banking conferences, and it is not hard to be convinced to return. The view is spectacular and the pillows are fluffy.

Also on the panel were a network equipment analyst who worked for Frank Quattrone at CSFB, and none other than Henry Blodget. It was nice to see Henry. He looked a little haggard, like he hadn't slept for a couple of months. He probably hadn't. Internet stocks had begun their long ride into oblivion back in March 2000 and I could just imagine his phone ringing off the hook with brokers in Sheboygan, Wisconsin, worrying about their shares of Infospace. Just three months earlier in February, I had sat through a presentation by Infospace CEO Naveen Jain who said, "I predict that Infospace will be the first company ever to have a one-trillion-dollar market capitalization." (By 2003, Infospace would be worth $300 million, 3/1000ths of the CEO's prediction.)

Now Henry had to deal with most of his stock recommendations heading towards zero. He was stoic. Metcalfe wanted to know when each of us thought the downturn would end. As I was out of the stock recommendation business, and didn't have any agenda to protect, I said what I thought. Remembering the summer of 1986, when semiconductor stocks had no bottom, I said that bottoms usually occur when the very last investor pukes his shares out, at any price. This got a good chuckle.

Henry had a problem in that he was still recommending a huge basket of stocks, like Yahoo and AOL but also Infospace, Pets.com and other Merrill Lynch banking clients. He said the downturn was almost over. He had to follow his recommendations or risk being called a liar.

So Metcalfe took a different tack.

"Henry, how can you still recommend shares of these companies that have no prospects of ever making any money?"

The ax syndrome whacked Henry as well.

"You've got to understand. If I stop recommending a stock, and the shares keep going up, there is hell to pay. Brokers call you up and yell at you for missing more of the upside. Bankers yell at you for messing up their relationships. There is just too much risk in not recommending these stocks."

I sat there stunned. This was exactly the opposite of my days as an analyst, where there was a risk in recommending a stock that might go down. Dennis Callahan's words "Don't recommend a stock that could go down," came flooding back. Henry had just said his philosophy not to stop recommending a stock that could still go up. What a bizarre twist. But of course, Henry was a bull market analyst. Until three months earlier, he had no experience recommending a stock that had gone down. He didn't know what it felt like.

This was the Ax Syndrome gone too far.

Wall Street was inverted. They yell if you *don't* recommend a stock. That could only be trouble. Henry had no way of knowing. He hadn't lived through a period when stocks went up *and* down. But plenty of people at Merrill Lynch had. Obviously he was out on a limb with nothing but a saw or a hangman's noose to get down.

. . .

In May 2000, *Business Week* ran a profile of Jack Grubman that set up his downfall. Jack said, "What used to be a conflict is now a synergy." There was his business model in ten concise words. He continued, "Someone like me who is banking-intensive would have been looked at disdainfully by the buy-side 15 years ago. Now they know I'm in the flow of what's going on."

Jack straddled the Chinese Wall, that conceptual separation of analysts and investment banking. It was Frank's "boutique within a bulge bracket" firm, but Jack himself was the boutique. The Chinese Wall ran through his brain, a bozo no-no at any Wall Street firm. The danger was that it would split him in two. In the end, that's exactly what happened.

Where was the compliance department at Salomon Brothers to manage this conflict? It was not like this problem hadn't arisen before. Firms knew how to manage it. Salomon and its parent company, Citigroup, looked the other way. It was too lucrative a straddle.

Jack made a huge blunder, and everyone on the Street caught him. A long time bear of AT&T, Jack hated the company. Hell, he made a career out of funding companies that took market share from AT&T. Citigroup's CEO Sandy Weill was on the board of AT&T, and AT&T's CEO Michael Armstrong on the board of Citigroup. Even with this, Jack stuck to his guns and was right.

Now AT&T was taking their Wireless division public, in

what would be a huge deal, spreading fees far and wide on Wall Street. Magically, Jack went from a bear to a bull on AT&T and recommended the shares. Voilà, Salomon was part of the AT&T Wireless deal. This raised more than eyebrows on Wall Street. The Chinese Wall had crumbled. Word spread of a Weill request that Jack "take a fresh look" at AT&T. Did Weill get to Jack, or did Jack get to Weill? Remember the "just break something" line to Margo Alexander at the room-destroying poker game? Involve management.

Either way, Jack knew Weill buttered Jack's toast. Of course, Weill knew Jack was a huge producer for Salomon/Citigroup. I'll even guess that fees from Jack's deals saved plenty of quarterly earnings reports by Citigroup. At the end of the day, Jack took one for the firm and became a hero. That is until both AT&T and AT&T Wireless blew up. Bad move, Jack.

In the end, it was the ax syndrome that got Jack. He kept recommending the shares of his companies all the way down. Worldcom, Teligent, Global Crossing, Qwest. He probably figured, correctly, that if he changed his ratings from Buys to Holds or Buys to Sells, the stocks would immediately go down. That would have screwed his clients. All of them. Joe Nacchio of Qwest would have called him up and cut off Salomon from future banking business. Lots of Jack's institutional clients who owned all these stocks would have called up and yelled at him with comments like, "You sold me these pieces of shit and now you're abandoning them? What kind of crap is that?"

Jack was stuck. These two constituencies boosted his ego, and his paycheck. Being right gets lost in the shuffle. When you swing a big ax, it's almost impossible not to cut your own head off.

. . .

I often ran into Henry Blodget at dinners that Pam Alexander hosted in New York and San Francisco. As things got rough, I told him that the only way to get through it was not to freeze up. Return clients' phone calls. Provide them with in-depth analysis. Keep working. OK, I was stealing the old "Bob Cornell recounting Ben Rosen freezing with a pile of pink phone message slips" story. I'm not bashful.

I called Mary Meeker to offer her the same advice. She didn't return my call.

But Henry stayed unfrozen. He never did pull his Buy recommendations, but he continued to be the ax in a number of stocks, including America Online. When they announced their merger with Time Warner, he was the most visible analyst. He blessed their $11 billion in cash flow prediction. Out of the frying pan into the fire. I should have told him to run away—that being an analyst was just a stepping stone to something else. And I should have told him the Keith Mullins document retention strategy story. Oh well, live and learn.

. . .

At the end of the day, Jack and Mary and Henry were right 51% of the time, just like they were supposed to be. Their problem was that they were right for two years, six months, and a day in a row, and then wrong for two years and six months in a row. You gottta mix it up. The long "right" streak made them superstars and the long "wrong" streak destroyed them.

CHAPTER 15

Spitzer Fixer

Dotcoms were like anchors with a six-month-long rope attached to them. At the other end of the rope about to be yanked into the abyss was telecom. By October 2000, telecom started giving up the ghost as well.

The stocks of Pets.com and Drugstore.com and other milk, er, Internet companies were going down as the ducks quacked in reverse and momos were selling. But telecom was a different story. They talked down their exposure to dotcom companies, and talked up their blue chip corporate customers. Worldcom, Global Crossing, and Qwest kept reporting decent revenues, and met analyst expectations. Their stocks were dropping, but not as fast as dotcoms, as telecom revenues provided a cushion to the downdraft.

It turned out that many telecom companies were just lying through their teeth. They employed the worst form of creative accounting—they were just swapping revenues with each other. "I'll buy $300 million of capacity from you if you buy $299 million of capacity from me. Investors will never know." Despite a fancy name, Indefeasible Rights of Use or IRUs, they were pure bullshit. These swaps were fluff used to distract investors from what was really going on: prices were collapsing and analyst expectations were way too high. At the time, Gary Winnick at Global Crossing and Joe Nacchio at Qwest were unloading shares. Bernie Ebbers at Worldcom, surprisingly, was not. Instead, he was borrowing against his Worldcom holdings.

Worldcom's deception went one step further than everyone else's. Their CFO Scott Sullivan started hiding costs by writing off network spending over years instead of expensing them right away, allegedly. This had the effect of making margins and cash flow look better, but the numbers were bogus. The accountants didn't pick up on this until it was too late. The Street and the press painted Jack Grubman as a scapegoat, for recommending Worldcom shares all the way down. I'm not letting Jack completely off the hook, but analysts have to believe management's numbers, and you can't hire your own accountants. You can snoop around to see where industry trends are headed, but when a company prints numbers, you have to trust them. In my view, Jack's bigger mistake was that he overestimated demand and pricing, versus being suckered by management.

It didn't matter. Jack's world started falling apart. His problems went beyond accounting. He had the whole industry wrong. He was the ax, and his stocks were getting slaughtered. I think I know exactly what was going through his mind—the dilemma that each day was a day too late to downgrade his recommendations from Buys to Sells. He froze. A downgrade would whack the shares even more. Ebbers would yell at him, as would Winnick and Nacchio. Banking business would disappear, but of course, it was going away anyway. Institutional investors would be relentless, shouting "asshole" when he picked up the phone and before they hung up.

Those damn ducks were screaming, "Get me out of these pieces of shit." But no liquidity meant every time a sell order hit, the stocks went down further. As their stocks collapsed, so did each company's ability to raise money, sorely needed to keep building their fiber networks or to pay interest on their huge debt. One by one, just about every one of Jack's clients (save Qwest which had merged with regulated phone company U.S. West) went under.

My own opinion is that none of the managers of these once successful companies set out to commit fraud. But as pricing and demand dropped, they did anything and everything to keep their stock up, to buy time to raise more money, or hope demand would return. By doing this, they crossed the line of honesty into fraud. It's a short walk.

• • •

By May 2001, Frank Quattrone was likewise in deep shit. His "boutique within a bulge bracket" firm had their fingers in a few too many things, and Frankie sat at the top of the org chart. He was accused of controlling analyst recommendations, doling out IPO shares to favored clients, and orchestrating kickbacks of IPO commissions. True? It didn't matter. The structure nailed him. It never should have gotten this far. CS First Boston, or any Wall Street firm, should have never allowed a boutique inside its structure. Chinese Walls are conceptual and busted in subtle ways. When a banker who gets 40% of revenue over some threshold is in charge of your bonus, you'll do whatever it takes to generate business. The pressure must have been intense.

My fund didn't get nearly enough shares of hot IPOs (that, of course, we deserved). I didn't give a rat's ass. Frankie made his own bed, now it was burning and people were throwing gasoline on it to put the fire out. Those shares that I didn't get were, allegedly (there's that word again), going to "Friends of Frank" accounts who in turn might steer investment banking business CSFB's way, and worse, to funds that would kick back some of the first day gains to CSFB by paying outsized commissions. In fact, CSFB would soon pay a $100 million fine to settle charges of these kickback schemes.

The *Wall Street Journal* ran a piece in May 2001 stating that Frank put undue pressure on analysts to put Buy ratings on stocks. The article quoted none other than Rick Ruvkun, claiming that back at Morgan Stanley, Frank pressured him to put a Buy on some stock. But little Rickie was a man, and stood up to him and put a Hold rating on it instead. This was a nice piece of revisionist history, in an attempt by Ruvkun to self-install a backbone.

So I sent an email to Frank (in Eliot Spitzer's archives now, no doubt):

> To: Frank Quattrone
> From: Andy Kessler
> It must have been Rick Ruvkun's shadow (the one he is still afraid of) that put out that Hold. Hang in there . . .
> Andy

I was trying to be nice. Hey, maybe I did give a rat's ass. Later that evening, I got an email back from Frank:

To: Andy Kessler
From: Frank Quattrone
How about writing an article about how you pressured Bill Brady into taking Mediavision public?

Now that was a low blow, but a well deserved one on my part. I was itching to do a multimedia deal back at Morgan Stanley, to prove to Frank that my research wasn't worthless. Brady and I wasted lots of time with Mediavision, and Frank smartly turned it down. This was well before management went to the pokey for fudging numbers.

Still, he had landed a solid jab. I remembered why, when anyone asked me if I worked with Frank Quattrone, I would reply, "I still have the scars in my back to prove it." I needed to return an upper cut. I was up half the night thinking it out. I sent back:

To: Frank Quattrone
From: Andy Kessler
It seems as if Brady hasn't turned down a deal since.

Of course, I was the least of Frank's problems. CS First Boston soon hired none other than John Mack, the former head of Morgan Stanley, who always liked bond traders more than he liked investment bankers. According to later press reports, in December 2000, Bill Brady sent out an internal email reminding people in the Technology Group about their document retention policy. This was perhaps a not-so-subtle reminder to get rid of everything. Frankie seemed to endorse this memo by reminding his people in another email that he sat through a grilling at the Miniscribe class action suit years before. These emails were smart and proper, if they hadn't been sent two days before an SEC investigation into CSFB's Technology Group. The boutique is busted. In early 2003, as these emails surfaced, Frankie was put on suspension. On March 4th, Frank resigned from CSFB.

. . .

As the bear market extended from 2000 all the way into 2003, investors, or should I say gamblers, were devastated. During the bull market, there was no compelling reason to buy Global Crossing shares. The actual story was lame. The company was losing billions, and while the swaps made it look better, real cash flow was negative. The company was bleeding to death in plain sight.

Investors just stopped analyzing. It got in the way of a great bull market. Instead, what had happened is everyone became momos, both funds and individuals. You bought Global Crossing because it was going up and the ducks were quacking. You might say you heard Jack Grubman recommending it, but unless you met him and walked through the story with him face to face, you couldn't blame Jack. You had to blame yourself. Gary Pilgrim showed the way, but the momo virus infected the entire market.

Al Harrison, the "state your conclusions upfront" portfolio manager at Alliance Capital in Minneapolis, was someone whom I respected a lot. He fell under the spell and chased Enron shares, another company employing creative accounting, all the way down.

Anyone who bought Global Crossing, or Drugstore.com, or Excite@Home, or Enron had a gambling problem. They were quacking, too. Investors Anonymous for everybody.

• • •

What was billed as the "Great Wall Street rip off" became big news. Investors saw their 401K retirement plans wiped out. Ex-employees of telecom companies saw their options and pension plans disappear. No one cried when dotcoms went down, probably because they were so stupid to begin with. Who would think of selling pet food online? But telecommunications—people thought—hey, I make phone calls and read my emails every day. Opinion quickly spread that these were standup companies raped by management and Wall Street.

A little-known Attorney General of New York, Eliot Spitzer, started getting involved. Under the guise of protecting individual investors, he pounced. But it was confusing. It was as if he had walked into a poker game with all the chips missing and three out of the five players shot dead, with no smoking gun to help figure out what went wrong.

Spitzer subpoenaed emails, from Salomon Brothers, from Merrill Lynch, and from everyone on the Street. Morgan Stanley was all too happy to oblige, but oh damn, they must have hit the delete button on those files, just as Ollie North had done with White House emails 15 years before.

Spitzer hit the jackpot with Merrill Lynch and Henry Blodget's emails. Henry had the audacity to be honest, saying that some of the companies he was recommending were POS, or pieces of shit. Well, no shit. The whole world soon believed that Wall Street research was corrupt. As the fan blades spun faster, and the POS hit the fan, Henry agreed to be part of an early retirement program that Merrill was offering to reduce headcount. He joined to huge fanfare, and in the end, slipped out the back door.

Mary Meeker, on the other hand, decided to stick it out. Several class action suits naming her were thrown out. There are probably plenty more where they came from. Mayree Clark, the director of research at Morgan Stanley, put out a press release saying, "Our research is thorough and objective, and Mary Meeker's integrity is beyond reproach." Something bothers me about this quote. It is unsatisfying. It got Mary off the hook, but has no meaning, almost as touchy-feely as her research.

• • •

In 2002, Worldcom's bankruptcy was even bigger news. Since Worldcom owned MCI, and everyone's friends and family used MCI, it landed with a huge thud. Finger pointing was nonstop. It was CEO Bernie Ebbers' fault. No, it was the CFO. No, it was the lax accountants. No, it was the lawyers. No, it seemed clear to everyone, it was Jack Grubman's fault. Why has nobody pointed at the momos, the ducks and the gamblers?

Mike Huckman, a reporter for CNBC, staked out Jack's apartment in New York, and started grilling Jack as he headed for work. It was great TV—a staged perp walk without handcuffs or prosecutors, simply a microphone, TV camera and viewers as judge and jury. Huckman was asking questions about Jack's stock recommendations. I saw a look in Jack's eyes that I recognized from year's past. Jack was very close to taking a swing at Huckman and probably would have knocked him out cold. I wanted to be there and hold Jack back, like in that elevator many years before. But Jack came to his senses, realized TV cameras were a great marketing tool, and actually started into his analyst marketing mode, pitching his view on telecom stocks. Jack was in the entertainment business, indeed.

Next thing you know, there is a Congressional hearing starring Bernie Ebbers, Scott Sullivan and Jack Grubman. I can't tell you how strange it is to see an old colleague raise his right hand and promise to tell the truth, under penalty of perjury. Jack was at his best. This was the ultimate marketing opportunity as an analyst! He couldn't deny that he worked closely with Worldcom, lest he lose his luster as a superstar. On the other hand, Worldcom was a ship that had already sunk. Jack had almost 20 years of training spinning client's questions into his own views and Congress was no match for him.

I called Jack a week later, to tell him to hang in there. He was amazingly upbeat, but hinted that there were other things to do in life than to be an analyst. He finally got the stepping-stone memo.

Citigroup CEO Sandy Weill apparently didn't like Jack's Congressional testimony or his lack of answers. Jack soon resigned, although it sure smelled to many like he was fired. Jack's emails began to surface and they were much juicier than Henry Blodget's. In emails to a client back in 2000, Jack the boaster wrote that he recommended AT&T shares to help Sandy Weill oust, or in Jack's words, "nuke" Citigroup then co-chairman, John Reed. In another, Jack said that he recommended the shares so Sandy Weill would help get his twins into the nursery school where Woody Allen's kid went.

This was perfect. It relieved him of the perception of whoring himself to get AT&T Wireless fees. After Spitzer released these emails to the press, who quickly wrote huge pieces on New York nursery schools, Jack retracted these statements by declaring he said these things to "bolster his professional importance." Now that's the Jack that I remember. Jack paid a $15 million fine and is now barred for life from the securities business. My guess is that is a relief to him. As a boxer, he knows how to get up from a knockout punch.

. . .

There are plenty of smoking guns to blame for the Internet and Telecom and Technology Bubble. None are very satisfying. Fed Chairman Alan Greenspan pumped the money supply to stave off a banking crisis based on Y2K computer problems and the excess money went into the stock market. Or how about excessive stock options led greedy management to fudge earnings numbers to pump up their stock. Yeah, maybe. It was structural problems on Wall Street that created the bubble, though excess money supply and corrupt management certainly contributed lots of the hot air.

Maybe Spitzer could fix all this, so I followed his moves with interest. Even though these were all my old buddies he was shitting on, I started rooting for Spitzer to uncover everything. The research problem, the "boutiques inside bulge bracket firms," no liquidity, breathing on stocks causing them to jump, missing Piranhas, momos, quacking ducks, lockups, gambling instead of investing, no place for retail investors to tread.

My mistake. I suppose Spitzer succeeded in exposing analyst dirt. But he turned Jack Grubman, Henry Blodget and Frank Quattrone (and to a lesser extent Mary Meeker) into the smoking guns that shot the poker players. He made it seem as if they were personally to blame for individuals losing money. That's a bit disingenuous. It seems that the structural issues were too complicated for Spitzer to deal with. Let's face it, you can't put a lockup agreement on TV.

There are still many structural problems. Regulators love to do things for the little guy, Joe Six-Stock. The SEC had passed a measure known as Reg FD or Regulation Fair Disclosure. Companies couldn't selectively tell analysts anything. They now had to broadcast their news far and wide, so that individual investors wouldn't be disadvantaged. This sounds innocent enough, except companies used it as a reason not to say anything ever to analysts, except once a quarter. Less news means more volatility, hurting Joe Six-Stock. How stupid.

In the wake of the Enron and Worldcom corporate and accounting scandals, Congress quickly passed the Sarbanes-Oxley bill, which forced management to swear that their results were true, imposing criminal penalties for false statements. As much as I would like to see a few lying CEOs do time up the river, all that the Sarbanes-Oxley will do is keep good managers out of the business. Why risk jail over a penny or two of earnings? Unintended consequences run amok.

Spitzer ended up negotiating a global settlement with the big players on Wall Street, who agreed to pay $1.5 billion in fines and make shallow changes to research in order to make Spitzer go away.

Too bad. I think that structural changes on Wall Street are happening anyway. The Virtual Morgan Stanley or the Synthetic Goldman Sachs is going to happen one way or another. Spitzer didn't have to do a thing beyond uncovering the dirt.

Institutional investors who got burned by Wall Street research were going to have to find some other sources to help them pick stocks anyway. When certain Wall Street analysts had blown their reputation, then you don't use them again. It's that simple.

What about the filters? Wall Street lost their Tom McDermotts and Richard Dickeys as their barking dog gatekeepers of the morning research meeting. No analyst could get airtime from them to pound the table on a stock selling at 150 times earnings. Hank Hermann and other piranhas no longer chewed up worthless analysts. And what about the quality filters that Morgan Stanley and Goldman Sachs had, things like two quarters of earnings and leader in their field before they would consider taking companies public? And what about the commitment committees? Are they still porous to let through skanky deals? Can the filters return?

They have to. Big, ugly Wall Street firms are hurt and now have to restructure around the new reality of Wall Street. Electronic trading, layers of independent research, and banking-only boutiques are here to stay. Unfortunately, the fines to make Spitzer go away will only freeze the field. It seems to me that Wall Street management reached into the pockets of their shareholders and paid big fines so they could keep the status quo. I have a bad feeling that Spitzer's "settlement" will merely perpetuate the old way of doing business much longer than its natural life. The structural changes and return of tough filters will take longer and be more painful to fulfill.

Is there some message to all this? Some note to future generations about how to avoid stock market bubbles, how to keep research honest, how to tame the cycles? Nah. They will learn it the hard way. Wall Street is a business. Analysts and salesmen and traders and bankers all make a living providing access to capital to businesses worldwide. For that task, the Street, as a group, gets to keep half of all the revenues they generate. But it's an information business. When you work on the Street, all you have is your reputation. Longevity comes from maintaining that reputation with all your constituents, including companies, institutional investors and Joe Six-Stock retail investors. Taint it, and someone else will fill your shoes. Creeping hubris is terminal.

Weird things will happen again, I guarantee it. Manias, frauds, axes, ducks—get used to them. If you try to legislate them away, you will end up killing the whole system, just as the Small Order Execution System, SOES, enacted in the name of fairness to small investors, killed liquidity.

Reputations trump legislation every day in my book. The tales of Jack and Frankie and Mary and Henry and all the rest are important, if only to show how powerful and then how fickle the Street can be. We really were just pieces of Wall Street Meat. Over time, Wall Street knows how to transform itself to stay a lucrative capital-raising machine. Those that abuse it won't last very long.

• • •

Jack Grubman is barred for life from the securities business and is probably too old to revive his boxing career. Frank Quattrone faces civil charges of improperly allocating IPO shares and controlling research. Mary Meeker is still an analyst, with integrity beyond reproach according to her boss. Henry Blodget quit Merrill Lynch and is back to being a journalist. The ducks have stopped quacking, but the class-action lawyer sharks are biting.

And me? I no longer have that wool tie, maroon shirt or double knit slacks. I'm glad, check that, grateful, that I got out of the research business when I did. I still get asked for stock tips, but get out of it by saying my Series 7 test has lapsed. That holds them off for a while. Without worrying about stocks, I sleep better, worry less when the market sells off, and my vacations aren't interrupted.

But I do care about Wall Street. It's not a casino, surreal or otherwise. It is a critical element to the success of the U.S. in the world markets. It is about access to capital for great companies, not stock tips and banking fees. I worry about the Spitzers and other outsiders ruining it for the sake of an idealistic view of fairness, or for the sake of keeping the status quo for Wall Street management. Modern Wall Street is a beast, constantly changing, always evolving, and fixing itself via the reputations made and lost every day. The Piranhas and the ducks and the axes all revolve around each other, and in the end, sort themselves out.

It was a wild ride, and I enjoyed every minute of it. I got out with at least some of my reputation intact, perhaps until someone reads this book. I may have to try that ride again someday.

Afterword

Half asleep from jet lag, I found myself sitting at the bar at the Mandarin Oriental in Hong Kong. I was stuck on this waste-of-time marketing trip in the early 1990s, meeting with money managers up and down the Pacific Rim. It sounds exotic, but it was just a plain pain in the ass. It was like the old Ohio Death March but on steroids, with me traveling to Tokyo, Seoul, Taiwan, Hong Kong, Singapore, and then back to New York in less than a week. I was supposed to meet a client from Hutchison-Whampoa or some such place at the bar and then be off to the Peking Duck House for a dinner of grizzle.

I was nursing a Tanqueray and tonic, and watching with amazement as the guy next to me downed four drinks in a row—Boodles, straight up, cut with a tiny lemon wedge. He looked familiar, like every other guy on a trading floor on Wall Street, dressed in a pinstriped suit, blue shirt with white collar, and a yellow paisley tie with matching braces.

"Bond trader, right?" I heard myself blurt out.

"Yep, about to go out to dinner with a salesman who owes me for all the business I've done. I need a little lubrication before dealing with a salesman. The name's Nick."

I laughed a little too loud and then asked, "Who you with?"

He told me Flemings or Schroeders or Barings, one of those dime-a-dozen British firms. Then he asked, "How about you?"

"I'm with Morgan Stanley, out of New York."

"Well, you're a long way from home. You trade govies?"

"No."

"Let me guess, converts?"

"No."

"Junk?"

"No."

"Not derivatives?" he asked suspiciously.

"No," I said almost embarrassed, "actually, I work in research."

"Oh, well, isn't *that* the civilized side of the business."

He chugged the last of his Boodles and took off.

To this day, I'm not sure if that was Nic Leeson, the "rouge trader" who tried to cover up some bad trades with more bad trades that eventually laid waste to Barings Bank. Who knows? Civilized indeed.

• • •

The civilized side of the business. There it was in a short phrase. Wall Street was a mosh pit, traders yelling across phone lines, salesmen hawking the latest deals, bankers scrambling to bring the next set of companies public. But research was, well, just civilized. I was not sure whether that was good or bad for me. It just was.

As an analyst, you weren't really part of the daily battles on Wall Street, the blocks of stock flying around, the price gyrations, the cold-calling of rich families, the fighting for banking deals. The yelling. The screaming.

You were above the scrum of the business. Buying, selling, trading, that was the dirty work of Wall Street, the machinery. Who cares how it's done, analysts thought, just get it done so you can pay me. Most traders kept chewable Pepto-Bismol at their trading turret; I had a Zagat's guide and a baseball schedule in my office. Yes, I spent lots of time on the trading floor—a zoo without cages—but was glad it was only a place to visit. I never really understood how these guys made money, and I didn't want to know. It was only later, when these traders and brokers and bankers stopped making enough money to pay for research, that analysts got their hands dirty and turned into bankers.

It may have been civilized, but I showed up every day, and worried myself sick. Did I know enough about Intel, or the PC business, or the Japanese, or some new market I hadn't thought of that would move the stocks in my group up or down? I worked inside a big, bustling company, but I was an entrepreneur, a one-man show. The market would judge me right or wrong. Daily moves could hurt or help, but I would be remembered for the big moves, the kind of market calls someone would remember 6 or 12 months later because they were so good.

If a stock recommendation worked, a client or salesman wouldn't call to congratulate me. But 6 months later they would phone up and just ask, "So what would you do here?" Over time, they knew if you were right or wrong. Your reputation was built over years. Your track record wasn't on paper, it was in the minds of all of your clients and all of your co-workers. You didn't have to tell them, they just knew.

But it got tiresome. You were always swimming upstream against the current of uncertainty. You were always second-guessing yourself. You always wanted to scream into the phone, "I told you to buy Intel 6 months ago at $15 and now it's $42 and now you're asking me what you should do? Go stick your head in the oven for not listening to me, you cretin." Somehow, it always came out slightly differently, Jim Mendelson-like: "As you know, I have been constructive on the group for some time, and while I maintain my buy rating, I would look to over weight on weakness." (Translation: I told you to buy it 6 months ago. It's now lower—I'm an idiot, but will never admit it publicly.) Whenever I caught myself saying that kind of garbage, I would leave my jacket out of sight on a hook on the back of my door, head to the elevator at 2:30 like I was going to get a cup of coffee, and go home.

Everyone on Wall Street gets these "What is this all about?" moments, or "Why am I doing this?" The deep types even ponder, "What does this all mean?"

On one hand, it means everything. Wall Street is where capitalist rubber meets the road. Without Wall Street, the wheels of capitalism, the economy, are up on chock blocks, and we are all serfs or communists or flipping burgers—or all three.

On the other hand, no matter what your specific job is, I can guarantee you that it doesn't mean anything. You are a cog, a pawn, a foot soldier. Traders on the New York Stock Exchange move pieces of paper around between owners. Analysts bust their ass picking stocks so the General Motors pension fund can do slightly better than the overall market. Salesmen push IPO deals for outsized commissions.

I was a piece of meat, everyone on the Street was too. The fat they rendered from us, like the grease that dripped off my Peking duck, was used as motor oil to keep the market slick and efficient. Everyone's position is lucrative but replaceable. This is not the kind of thinking that helps personal well-being and self-worth.

But so what? Everybody on the Street is a cynic. You're getting paid plenty. Got a problem? Hire a therapist like everyone else.

. . .

Of course, not everyone knew the code of Wall Street. Not everyone was hip to the cynicism. Individual investors got the shaft in all of this. No one told them about the subtle shifts that were taking place. "What do you mean an analyst who works with bankers can't be trusted?" the ordinary investor might have asked. "I never read that in *Investors Business Daily* and they know everything."

So, when the bottom fell out, it was no wonder analysts and bankers became the poster children for the bubble. It was false advertising. Analysts weren't analyzing. Bankers were just interested in fees. Individual investors only knew of reputations from CNBC, not from the day-to-day and year-to-year grind of investing.

For those of us in on the game, your reputation was everything, because at the end of the day, that is all you have on Wall Street. Caring about your reputation, that's human nature. But on Wall Street, it's the key to longevity.

As an analyst, you have to have a thick skin. It's hard, because also like human nature, no one really has a thick skin. Being wrong, even 49% of the time, hurts like hell. It stings and stings. "Civilized," my ass. The trick was to turn your pain into your advantage.

For me, I didn't have much other choice. A few of my competitors had moles, friends inside companies that would tell them how the quarter was shaping up. They would hear that May was a good month and pound the table on the stock, not citing their mole but instead "industry contacts in the sales channel." And the stock would go up. I found this a bit sleazy, but no one else did. I suppose I was just jealous. I didn't have any moles whispering in my ear. Other competitors would just make stuff up to justify their ratings. Clients would call me and ask what I thought about these statements, and it took me years to get the courage to say that my competitor just made it up. With no moles and a limited imagination, I had to figure out where the world was going myself, think long term, and give up the short-term moves to my squirrely competition.

Bob Cornell, my first boss and mentor, taught me to be early, but not too early, which confused me for the longest time. What he meant is that often you have to be wrong to be right. You have to say buy as a stock is going down, and then work your long-term arguments with the salesforce and institutional clients, in the face of being dead wrong every day for months on end. That's how to make people money—get them into a stock when everyone is fleeing. That's what they remember you for. That's how you build an iron-clad reputation on Wall Street, whether you are in civilized research or anywhere else. But you can imagine the pain of being wrong for months before being proved right. This happened to me with my first sell call and my first buy call. Not fun! And then when you are finally right, and stocks start to move in your direction, all your competitors quickly flip and start recommending the stock on the way up. But the real clients know you were early and right.

Unfortunately, there is no way to cut corners. You can tell clients how great you are, but they know the truth. They decide your reputation, not you. When they turn on you, and your reputation turns sour, there's no one around to let you know. They just snicker behind your back. Your reputation is your career. You can't buy it, you can't steal it, you can't cheat to keep a reputation, not in the long run anyway.

But it is really easy to piss it away. Integrity and reputation are inextricably linked, like handcuffs. I have to admit, I am not big on the integrity thing, mostly because I never quite thought about it. This isn't some big morals lesson. You have your integrity until you are confronted with a situation in which you can trade it away for something better, like *money*. There were always ways to do this, if you wanted. I just never found any that were worth it. Not that I actively looked, but they come up every day.

It is so easy to screw other people on Wall Street, it's not even funny. But you can only screw them *once*.

Every trader knows that you can call up a client offering to sell them 10,000 shares of Merck.

"Any more shares behind that?"

"No, this is the clean-up print."

"Great, I'll take them at 52½."

"Done."

Ten minutes later, a block of 200,000 shares trades at $51¼.

Ouch. That trader, maybe even the trader's firm, is cut off, for a week, for a month, forever.

For salesmen and for traders, there are no contracts, no legal recourse for bad trades or bad deals. There are not even handshakes. Everything is done by phone. You cut corners and lie to get a trade done, and you're not going to be doing too many more. You have to have integrity to survive your first week on the job, and then keep it for your entire career.

In research and banking, things aren't so clear-cut.

I would often get calls from salesmen saying, "I get a sense you aren't as positive as you once were on the group. Tell me before you cut your ratings so I can get my clients out." Yeah right, you'll be my first call, but your clients won't respect me in the morning.

Bankers can do lots of deals, but if a few too many of them blow up down the road, buyers of those deals have long memories.

An analyst has too many constituents to favor any one of them—salesmen, traders, institutional clients, brokers, individual investors, bankers, the press, friends and family who want stock tips, let alone cab drivers in bull markets who see your Morgan Stanley bag and pepper you for stock tips. It is not hard to handle and balance all of these folks. You just weigh them based on how much they are paying you. The broker from Moline chewing you out for a bad stock call is pretty easy to dismiss, you just hold the phone up in the air until the noise stops coming out and then say "It won't happen again." Then hang up. The guy from Fidelity questioning which part of your anatomy contains your brains is a little tougher to dismiss. You just pray that he is pissed off because his fund's numbers are awful and will probably get canned at the end of the year, and you can start fresh with some new snot-faced MBA graduate in January.

But the biggest risk to one's integrity and therefore to one's hard-won reputation is the sweet siren of investment banking. Intellectually, investment banking is true civilization. The stock market is about access to capital, and investment banking is about providing capital to companies deemed worthy by the market. It is almost altruistic charitable work, Mother Teresa in a pinstriped suit. Investment bankers float from company to company, offering to apply money toward a worthy cause—"Ignore those onerous fees, that is just something we need to charge to cover our overhead."

To analysts, banking is nirvana. It has the feel-good characteristics so often missing from an analyst's life. It is the yin to the yang of getting yelled at by that know-nothing client at Fido. You feel personally responsible for a company's success—"I made Amazon what they are today, by getting them the capital they need."

"You did get paid for that, right?"

"Well, yes, but did I mention the whole rubber-meets-the-road, wheels-of-capital thing? Oh, never mind."

. . .

Analysts work with investment bankers because investment bankers pay more than traders, salesmen, and brokers. The feel-good-about-it stuff is Rationalization with a capital R. On the surface, there is nothing wrong with this; bankers need analysts to analyze the prospects of an industry or a specific company. But in reality, bankers extract a price. Integrity, even if it is only a small, infinitesimal piece of integrity, is sacrificed to get a deal done. I did it; every analyst does it. And once you lose even a tiny piece of integrity, it is a slippery slope, a greased gangplank to losing your reputation.

You don't mean to throw it all away, it's just the balance between all your constituencies gets out of whack. When you take a company public or help raise $1 billion in capital after they are already public, you end up working for the CEO of the company. Institutional clients get hosed, so do brokers, individual investors, salesman, and traders. Your firm stands to clear $10 million in fees from the deal, and you might get a mill or two of that, but the rest of the Street notes the loss of reputation.

Jack Grubman took it to the extreme. He was squarely in the CEO's camp, but convinced investors to come along for the ride because of the revolution taking place in telecom. His reputation suffered, and then everybody turned on him when things went sour.

Mary Meeker may think investors were impressed that she was involved with the Netscape IPO, but clients recognized the shift in her job function from analyst to cheerleader, and her reputation suffered accordingly.

Frank Quattrone did so many deals—a lot of good ones, but a few too many stinkers—that investors became leery of CS First Boston IPOs. They would still play the game, hoping to get shares that would immediately go up in hot deals, but my sense is that most institutions weren't interested in Quattrone's clients as long-term holdings.

But it wasn't just these individuals. Entire Wall Street firms forgot that their reputations were at stake when they did deals and touted stocks. Investors have long memories, and the name in flashing lights at the top of the building is the one that is remembered most.

. . .

The biggest question I get from friends, acquaintances, and myself is why did I get out? I sat next to or worked with all the notorious Wall Street players and got out with my reputation intact. I must have done something right. What did I see about Wall Street research that made me run away? Was my integrity so important that I left a lucrative position to maintain my reputation?

I wish I could point to upstanding citizenship, strong values, and a sense of right and wrong. But that's a load. The real answer is dumb luck. When I left Morgan Stanley, a salesman and friend David Boucher called me up and yelled at me, "Andy, don't be stupid. You're number two in *Institutional Investor* and a rising star around here. I know it sucks working here. People yell at you all the time, management by fire. But you've got to stick it out. With all the deals to be done, pretty soon analysts are going to be paid a million a year, maybe multiples of that. Just hold your nose and deal with it."

To be honest, I'm just as greedy as the next guy. I just wanted a more interesting role. I didn't want to analyze companies, I wanted to own them, and have some other clown analyze them. Analysts follow companies, I wanted to lead them. OK, not run them, that would be too much work, but be ahead of trends. I had missed a few big wealth cycles, semiconductors, PCs, and software happened in my post-collegiate years, and I was getting paid as an analyst to figure out which of many existing companies would do well, rather than help invent a whole new set of companies.

When the dumb-luck answer isn't sufficient to the why-did-you-get-out question, I say with a straight face that to this day I thank Quattrone for getting me out of the business. He was the best in the business, but didn't think twice about screwing me, stabbing me in the back, and going to my boss in an effort to cut my bonus for not being a "team player." To be fair, that was his job—to get analysts to help bring in deals. For me, it wasn't an integrity thing, I just hated the companies that I was good at following and was looking at more interesting pastures. That didn't sell well with Quattrone who had to generate real revenues. Don't get me wrong. I like him. I wish him well and hope he beats the rap, but I still have scars from the backstabbing that make me wince when the weather changes.

It's odd that Quattrone has ended up as the bad guy, maybe even the fall guy for the Internet bubble. He is accused of conspiracy to destroy documents, or some such charge, for the wording and timing of an email he sent out. I have no clue as to his innocence or guilt, but I do feel bad for what he is going through. Integrity and reputation are critical on Wall Street, and he seems to be taking the fall for everyone whose integrity slipped. Fair or not, it points out both how fickle Wall Street can be and how important it is to guard one's reputation more than anything else.

• • •

I've heard from a lot of folks since writing this book. Some agreeing with me; others calling me a buffoon. I don't think this book hurt or helped anyone's reputation on Wall Street. Reputations are made or lost in the market. The participants in the wild-and-woolly stock market just know. Plain and simple. Wall Street is not Hollywood or Madison Avenue. You are right or wrong, honest or dishonest over a period of time. You can't wave away what everyone thinks of you, no matter what you do. You lie or cheat or give bad advice, and there is almost no redemption. If it sounds like I hold grudges, I don't. I just called it like I remembered it. Nothing I've written is not already imprinted in stock market history. Perhaps I just brought it to the surface to be batted around a bit.

Most things haven't changed. Goldman and Merill and Morgan still fight it out for deals. Research analysts aren't allowed to participate, one of the demands of Eliot Spitzer's $1.4 billion settlement. But from what I understand, analysts still are bankers. Of course they are. The problems with trading commissions and thin spreads hasn't changed, and bankers are the only upside in analysts' compensation. Meanwhile, the buyside (which can still say "asshole" before hanging up the phone) needs research and any and all help it can get finding companies that will generate great long-term returns. They hire their own, hire outside firms, pay Wall Street when they can, but at the end of the day, they are still underserved. It spells opportunity.

• • •

Would I do it again? If I were 25 years old, would I go through the struggle of gaining and keeping a reputation on Wall Street again, as an analyst or in any other role? Yes, in a heartbeat. There is no better vantage point to watch and participate in this thing called capitalism. It's not about clipping coupons or flipping shares around anymore; it is about creating new instruments to fund growing companies. Banks are increasingly obsolete. No one goes to a bank for a loan anymore. Investment banks are the gateway to forming capital in America.

When Intel inevitably misses its earnings forecast every couple of quarters and the stock is down 3 on huge volume and CNBC is buzzing with analysts slashing numbers, feel sorry for the poor analyst who has to cover this roller coaster of an industry. In the past, an event like this would not only ruin my day, but the next 6 months of scratching and clawing to rebuild my reputation as an analyst who knows what is going on and can actually predict Intel's earnings. Today, I laugh out loud, have another sip of coffee, and move onto more pressing matters. But it doesn't mean I wouldn't do it all over again in a flash.

When I joined Paine Webber as an analyst, I was incredibly naïve. But that naïvety helped me take risks and make nonconsensus calls that worked out. If I was older and more experienced, I might have not been willing to take those risks, and perhaps would never have been as successful. When I left Wall Street, I was wiser, but honestly, still pretty naïve about how things worked. One thing I do know, however, if you think about risks too much, you'll never take them. If you worry about your reputation, you'll never get one. If you nibble away at your integrity to win any amount of success, you've blown it all. Failure, at least 49% of the time, is tolerated. I made my fair share of big, stupid boneheaded mistakes. But over the long run, through no real plan or design, just in keeping my head down and in trying to be smarter than anyone else, I kept my integrity and reputation and lived to tell my story. Civilized? I'm not so sure.

Index

Index